VEST POCKET
REFERENCE BOOK

Revised Edition

OTHER BOOKS BY THE AUTHOR

Architectural Drafting, 5th Edition

Mechanical Drafting

Blueprint Reading—An interpretation of Architectural Working Drawings

Materials and Methods for Contemporary Construction

Estimating Building Construction Quantity Surveying

Construction: Systems and Materials

BUILDERS VEST POCKET REFERENCE BOOK

Revised Edition

Compiled By

WILLIAM J. HORNUNG

Formerly
Director of Training
National Technical Institute

Instructor
New York University
Division of General Education

Estimating Instructor
Institute of Design and Construction

Founder and Director
Long Island Technical School

PRENTICE HALL PRESS • NEW YORK

Copyright © 1955, 1975 by Prentice-Hall, Inc.

All rights reserved, including the right of reproduction
in whole or in part in any form.
Published in 1986 by Prentice Hall Press
A Division of Simon & Schuster, Inc.
Gulf + Western Building
One Gulf + Western Plaza
New York, NY 10023

Originally published by Prentice-Hall, Inc.

PRENTICE HALL PRESS is a trademark of Simon & Schuster, Inc.

Library of Congress Catalog Card Number: 55-7137
ISBN 0-13-085944-3

Manufactured in the United States of America

40 39 38 37

First Prentice Hall Press Edition

TABLE OF CONTENTS

BUILDERS
VEST POCKET
REFERENCE BOOK

INTRODUCTION

Building contracting requires a complete knowledge of all phases of construction work. The estimator in a well-organized contractor's office is, in most cases, the office's center of activity. He usually controls the job from the time the first contact is being made, concerning a proposed building, up to the time the job is well under way. His duties include quantity surveying, interviewing subcontractors, getting quotations on materials, and preparing the estimate.

The preliminary estimate is usually being made before plans and specifications are complete. Such an estimate enables the architect and engineer to find out the extent to which he may go in developing and in detailing final plans and specifications.

Regular estimates are made up by the general contractor's estimator, while preliminary estimates are usually made up in the architect's office.

When construction gets under way, the estimator will turn over complete plans, specifications, copies of all subcontracts, and all other available data to the superintendent, who from then on is in complete charge of the job. If all estimator's records are complete at this time, at a later time nobody will be able to hold him responsible for any mistakes or negligence.

The architect makes the plans, writes the specifications, and takes care of the owner's interests at all times. In the beginning the architect deals with the owner only. After drawings and specifications have been prepared, usually the architect, or sometimes the owner, will invite several contractors to bid on the job. Public work must be advertised, the bids submitted up to a certain date, and the lowest bidder must be given the job. The architect will lend one set of draw-

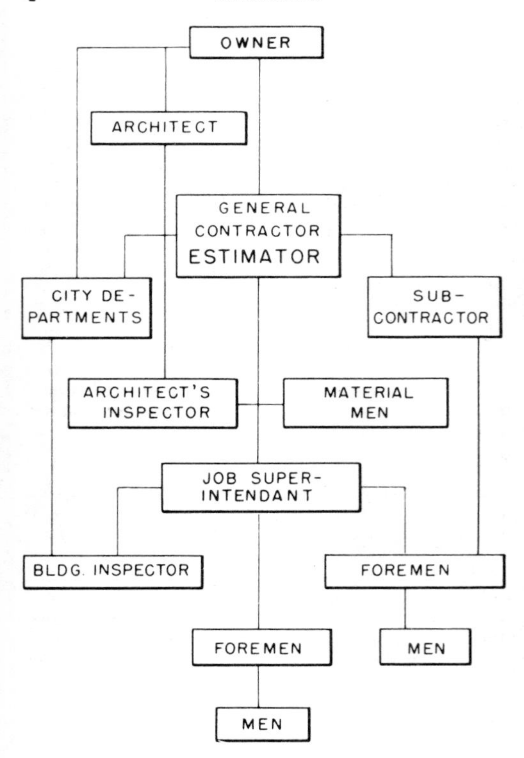

ings and specifications to each of the bidding contractors, and the contractors will prepare their estimates based on these data.

"Subcontractors" are men or firms who work for general contractors and who handle those parts of the building which the general contractor either does not want to or cannot handle himself. It is the general contractor's duty to co-ordinate the work of all subcontractors and to see to it that every "sub" is on the job at the proper time and that the various trades do not interfere with each other. The general contractor sometimes has the carpentry or the brickwork done by his own workmen; sometimes he is just the co-ordinator of all the trades.

"Materialmen" applies to firms selling the materials. For instance, the lumber dealer is a materialman. He sends the lumber to the job but does not do the carpentry work.

"Mechanics" are the men who work with tools. Some trades employ apprentices and helpers. These may also be classified as laborers.

"Specifications" guide the whole construction job as to quality of materials, workmanship, and relations between the parties concerned with the job.

A good specification should be written in the same sequence as the various trades start working on the job, thus starting with the general conditions and continuing with demolition, clearing site, excavation, etc.

The "contractor," sometimes called "general" or "builder," is responsible for the entire job. He has to obtain building permits and establish the necessary safeguards, temporary toilet facilities, and water supply.

ESTIMATING CHECK LIST

Before submitting bid, check estimate carefully with this list and specifications to avoid possible mistakes and omissions.

GENERAL CONDITIONS:

Superintendant, timekeeper, material clerk, foremen, watchmen, ladders, stairways, water closets, heat and light, temporary enclosures, barricades, partitions or screens, removing and replacing water hydrants, telegraph and telephone poles and wires, sewers and drains, cleaning floors and windows, removing rubbish, surveys, photographs, blueprints, job office supplies, traveling expenses.

PERMITS, INSURANCE, BONDS, TAXES:

Building, sewer, water, and street permits. Workmen's compensation or employer's liability insurance. Public liability insurance, owner's contingent or contractural insurance, contractor's contingent insurance, fire and/or tornado insurance, contractor's equipment floater insurance. Old age benefit and unemployment compensation insurance, sales taxes, surety bond, street obstruction bond. Labor, material, maintenance, supply and performance bonds.

CONSTRUCTION PLANT, TOOLS AND EQUIPMENT:

Power shovels, bulldozers, concrete mixers, truckmixers, hoists, pumps, air compressors, gasoline engines, power saws, derricks, hoisting towers, concrete buckets, hoppers and spouts, motor trucks, conveyors, trenching machines, wheelbarrows, concrete carts, wood and metal scaffolding, mortar mixers, salamanders, water hose, buckets, picks, shovels, sledges, miscellaneous small tools, temporary job office, sheds and tool houses.

WRECKING AND CLEARING SITE:

Removing old trees and stumps, clearing underbrush, wrecking old buildings, foundations, and retaining walls. Removing old material from the premises.

EXCAVATING, BACKFILLING, AND PUMPING:

Basement, subbasement, piers, footings, pits and trench excavation, backfilling, grading and filling, blasting, pumping.

SHORING AND UNDERPINNING:

Shoring and underpinning adjoining buildings, streets, and alleys.

PILINGS OR CAISSONS:

Wood or concrete piles, wood or steel sheet piling. Caissons, including excavating, iron rings, and concrete.

FOUNDATIONS AND RETAINING WALLS:

Footings, foundations, and retaining walls of concrete, rubble, or cut stone, sewer, common, or paving bricks, concrete blocks or vitrified tile.

WATER- AND DAMPPROOFING:

Integral or membrane waterproofing, waterproof paints or plaster coats for basement or foundation walls, boiler rooms, sump pits, etc. Dampproof paints or compounds for interior or exterior brick, tile, or concrete walls. Waterproofing cement or concrete floors. Concrete curing compounds.

CEMENT FLOORS, WALKS, OR PAVEMENTS:

Sidewalks, basement floors, porch floors, areaways, steps, driveways, cinder or nailing concrete, concrete floor fill, roof saddles, cement-finish floors and base. Curing compounds.

BRICK, CONCRETE, AND GLASS MASONRY:

Face or press brick, enamel or glazed brick, firebrick, common brick, sewer brick, bricking-in boilers, incinerators, catch basins, man holes, brick stacks, wall capping, flue lining, brick mantels, firebrick lining, hearths and backhearths. Concrete blocks and backup blocks, glass blocks, drain tile, mortar, mortar color, wall ties, cleaning and pointing.

CAST STONE, CUT STONE, OR GRANITE:

Cost of cast stone, cut stone, or granite at quarry or mills, freight, cartage, cutting, carving, lewising, metal anchors, scaffolding, labor handling and setting, mortar, back painting or back plastering, cleaning and pointing.

TERRA COTTA:

Cost of interior and exterior terra cotta at factory, freight, cartage, metal anchors, filling with concrete, scaffolding, hoisting, labor handling and setting, mortar, cleaning and pointing.

ARCHITECTURAL CONCRETE:

Cost of details, models, waste molds, erecting, removing, cleaning and painting forms, form ties and spreaders, reinforcing steel, mixing and placing concrete, patching and rubbing.

REINFORCED CONCRETE:

Wood or metal forms for concrete floors, stairs, beams, girders, columns, etc. Labor erecting, removing, and cleaning forms. Reinforcing steel, freight, cartage, cutting and bending, handling, placing, and wiring in place. Clay, gypsum, or metal floor tile. Labor handling, hoisting, and placing tile. Cement, sand, gravel, slag, cinders, or crushed stone. Labor handling materials, mixing, hoisting, and placing concrete. Winter protection against freezing. Concrete curing compounds.

TILE, GYPSUM, OR CONCRETE MASONRY:

Tile floor arches. Clay tile, gypsum, or concrete masonry partitions and wall furring. Beam, girder, and column fireproofing. Structural tile bearing walls and backup tile. Tile floor fill and furring. Book tile. Structural roof tile. Freight, cartage, mortar, labor.

ROUGH CARPENTRY:

Roof trusses. Framing and erecting timbers for columns, beams, girders, and heavy joists, roof framing, laminated floors, heavy factory floors, rough subflooring, wall and roof sheathing, roof saddles, stud partitions, wood furring strips, wood grounds and window blocking, rough stairs, bevel or drop siding, shingles, hook strips, shelves, etc.

FINISH CARPENTRY:

Door and window frames, sash, exterior and interior doors, door jambs, door and window trim and casings, wood base, chair rail, picture mold, panel strips, cornice, wood paneling, wardrobes, cases and cabinets, linen cases, ironing boards, mantel fronts and shelves, heavy garage and factory doors and frames, outside blinds, porch materials, door and window screens, storm doors and sash, labor handling and erecting. Wood stairs.

WOOD FLOORS:

Softwood, hardwood, or parquetry flooring, plank flooring, wood-block flooring, etc. Labor handling, laying, scraping, planing, sanding, painters' finish, etc. Nails or mastic. Paper protection for floors.

INSULATION, WALLBOARD, SOUND DEADING, ACOUSTICAL TILE:

Deading felt, quilt or building paper between floors, insulation for walls and ceilings. Wood furring or nailing strips between studs or ceiling joists, vapor-seal sheathing, insulating sheathing, wallboard, plank or tile, corkboard, gypsum wallboard, hard board, accoustical tile. Metal, wood, or fiber moldings. Nails and mastic. Labor handling and placing.

WEATHER STRIPS AND CALKING:

Calking door and window frames, weather strips for doors and windows. Interlocking thresholds.

LATHING AND PLASTERING:

Wood, metal, or gypsum lath. Metal channels, rods, and lath, staples and labor for ceilings, partitions, and furring. Metal door and window trim. Corner bead, base bead, etc. Labor handling and installing. Plain interior plastering, Keene's and Portland cement base and wainscot, ceiling and wall coves. Imitation Caen stone. Ornamental plaster cornices, beams, columns, bases, caps, wall and ceiling panels. Exterior plaster or stucco.

FIRE DOORS AND WINDOWS:

Hollow metal door and window frames, sash, and doors. Kalamein door frames and doors, casings, hardware, wire glass, freight, cartage, labor.

HOLLOW METAL DOORS, TRIM, AND FURNITURE:

Art metal frames, doors, and windows. Elevator doors, hangers, track, and hardware. Jambs for doors and borrowed lights, transoms, trim or casings, base, chair rail, picture mold, etc. Desks, lockers, filing cabinets, chairs, hardware, freight, cartage, labor.

STEEL SASH, DOORS, PARTITIONS, AND SKYLIGHTS:

Steel sash, doors, partitions, saw-tooth and monitor sash, skylights, rolling steel shutters, glass, putty, freight, cartage, labor.

SHEET-METAL WORK:

Skylights, ventilators, flashings, gutters and downspouts, metal hips, valleys and ridge roll, metal cornices, ceilings, steel siding, tin, steel, lead, and copper roofing. Glass, putty, painting, freight, cartage, labor.

ROOFING:

Wood, asphalt, or asbestos shingles. Slate or tile. Ridge, hips and valleys. Prepared roll roofing, built-up roofing. Corrugated metal or asbestos roofing and siding. Nails, clips, freight, cartage, labor.

TILE AND MOSAIC:

Ceramic, mosaic, and quarry tile floors, base, and stairs. Terrazzo floors, base, stairs, and wainscoting. Glazed or enamel base, wainscoting, and cap. Built-in fixtures. Tile mantels, hearths, and backhearths. Cement fill under floors. Freight, cartage, labor.

ASPHALT, CORK, LINOLEUM, RUBBER AND COMPOSITION TILE:

Asphalt, cork, linoleum, rubber, and composition tile, base, and wainscoting. Felt, cement, freight, cartage, labor.

ART MARBLE AND SCAGLIOLA:

Art-marble floor tile, door and window trim, wainscoting, stair treads and risers, wall panels, columns, bases, and caps. Freight, cartage, labor.

MARBLE AND SLATE:

Marble floor tile, base, and wainscoting, partitions, treads and risers, handrail, mantels, door and window trim. Freight, cartage, erection. Slate partitions, urinals, floor slabs, floor tile, laundry tubs, treads and risers, freight, cartage, erection.

GLASS AND GLAZING, STRUCTURAL GLASS:

Window and plate glass, mirrors, leaded or other art glass, plain or ribbed wire glass, obscure, prism, and fancy glass. Structural glass, glass blocks. Putty, freight, cartage, labor.

PAINTING AND DECORATING:

Painting exterior frames, wood- and metalwork. Painting, enameling, varnishing, or waxing interior woodwork, base, chair rail, picture mold, cornices, door and window trim, cases and cabinets, stairs. Finishing wood floors, painting and tinting walls and ceilings, casein paint, cold-water paint, whitewashing, wallpaper and paper hanging, canvas, Sanitas, etc.

STRUCTURAL IRON AND STEEL:

Iron castings, grillage, structural columns, girders, beams, roof trusses, crane tracks, bolts, rivets, freight, cartage, unloading, handling and erecting, riveting, welding, painting. Cast-iron columns, bases, steel stacks, junior I beams, erection, painting.

MISCELLANEOUS IRON AND STEEL:

Area gratings, manhole covers, coal chutes, ash traps, cleanout doors, bumpers, bases and caps for wood columns, joist hangers, anchors, straps, garbage and package receivers, window guards, pipe railings, sidewalk doors, vault doors, wall safes, concrete inserts, etc.

ORNAMENTAL IRON, ALUMINUM, BRONZE, AND STAINLESS STEEL:

Iron stairs, fire escapes, elevator doors and enclosures, ornamental posts, lamps and lanterns, doors and frames, standards, tablets, partitions, railings, gates, grilles, signs, thresholds, push plates, metal store fronts, freight, cartage, labor.

ROUGH HARDWARE:

Nails, screws, bolts, anchors, sash cord or chain, sash weights, pulleys or balances, window fixtures, expansion bolts, garage and sliding door tracks and hangers, etc.

FINISH HARDWARE:

Butts, door locks, escutcheon plates, push plates, door holders, door checks, window lifts, locks, burglar locks, cremone bolts, casement-window adjusters and fastners, French-door hardware, case and cabinet hardware.

PLUMBING, SEWERAGE, AND GAS FITTING:

Sewer excavation, sewer and soil pipe, pipe and fittings, pipe covering, hot- and cold-water supply, fixtures such as bathtubs, showers, sinks, lavatories, water closets, laundry tubs, drinking fountains. Gas pipe and fittings, water tanks and heaters, water softener, gas range and stoves, freight, cartage, labor.

VACUUM-CLEANING SYSTEM:

Pipe and fittings, electrical outlets and connections, motors, hose and appliances, freight, cartage, labor.

HEATING AND VENTILATING:

Steam or hot-water boilers, pipe and fittings, radiators, valves and air vents, pipe covering, painting, freight, cartage, labor. Thermostat. Warm-air furnaces, blowers, humidifiers, pipe and fittings, ducts, registers, thermostat, smoke pipe. Freight, cartage, labor, oil burners, tanks, stokers.

AIR CONDITIONING:

Air washers, humidifiers, ventilators, fans, metal ducts, refrigerating equipment, motors, controls, freight, cartage, labor.

POWER-PLANT EQUIPMENT:

Boilers, engines, dynamos, transformers, pumps, switchboards, cranes and conveying machinery, refrigerating machinery, stokers, machine foundations, pipe covering, freight, cartage, labor.

ELECTRIC AND POWER WIRING:

Outside service, conduit, outlet boxes, floor boxes, wire, switches, base plugs, bracket outlets, radio outlets, service boxes, speaking tubes, watchman's clocks, transformers, freight, cartage, labor.

LIGHTING FIXTURES:

Wall and ceiling fixtures and lamps, freight, cartage, labor.

ELEVATORS, ESCALATORS, DUMB-WAITERS:

Sidewalk lifts, dumb-waiters, escalators, freight and passenger elevators, tracks, doors, hangers, enclosures, hardware, signal devices, freight, cartage, labor.

AUTOMATIC SPRINKLER SYSTEM:

Pipe and fittings, tanks, pumps, motors, engines, city-main connections, sprinkler heads, valves, freight, cartage, labor.

MAIL CHUTE:

Mail chute installed in building.

MISCELLANEOUS EQUIPMENT:

Kitchen cases, gas or electric ranges, refrigerators, dishwashing machines, ventilating fans, window shades, venetian blinds, curtain rods and brackets, radiator covers and humidifiers, garbage cans, incinerators, wall beds, hall and stair carpets, pneumatic tubes, and other mechanical devices.

SUMMARY OF ESTIMATE

Building	Location	Estimate No.
Architect	Owner	Date
Cubical Contents	Cost per Cu. Ft.	Estimator
Floor Area, Sq. Ft.	Cost per Sq. Ft.	Checker

1. General conditions and overhead expense				
2. Permits, insurance, bonds, and taxes				
3. Construction plant, tools, and equipment				
4. Wrecking and clearing site				
5. Excavating and pumping				
6. Shoring and underpinning				
7. Piling or caissons				
8. Foundations and retaining walls				
9. Water- and dampproofing				
10. Cement floors, walks, and pavements				
11. Brick, tile, and concrete masonry				
12. Cast stone, cut stone, or granite				
13. Terra cotta				
14. Architectural concrete				
15. Reinforced concrete				
16. Tile, gypsum, or concrete-block fireproofing				
17. Rough carpentry				

SUMMARY OF ESTIMATE (*Continued*)

Building	Location	Estimate No.
Architect	Owner	Date
Cubical Contents	Cost per Cu. Ft.	Estimator
Floor Area, Sq. Ft.	Cost per Sq. Ft.	Checker

18. Finish carpentry				
19. Wood floors				
20. Insulation, sound deadening, accoustical tile				
21. Weather strips and calking				
22. Lathing and plastering				
23. Fire doors and windows				
24. Hollow metal doors and trim				
25. Steel sash, doors, partitions, skylights				
26. Sheet-metal work, skylights, flashing, etc.				
27. Roofing, built-up, tile, slate, metal				
28. Tile and mosaic floors, walls, stairs				
29. Asphalt, cork, linoleum, and rubber tile				
30. Art marble and scagliola				
31. Marble and slate				
32. Glass and glazing, structural glass				
33. Painting and decorating				
34. Structural iron and steel				
35. Miscellaneous iron and steel				
36. Ornamental iron, aluminum, bronze, steel				
37. Rough hardware				
38. Finish hardware				
39. Plumbing, sewage, and gas fitting				
40. Vacuum-cleaning system				
41. Heating and ventilating				
42. Air conditioning				
43. Power-plant equipment				
44. Electric and power wiring				
45. Lighting fixtures				
46. Elevators, escalators, dumb-waiters				
47. Automatic sprinkler system				
48. Mail chute				
49.				
50.				
51.				
52.				
53.				
54.				
55.				
56.				
57. Total cost				
58. Profit				
59. Surety bond				
60. Amount of bid				

ACCURATE EARTH EXCAVATION

GENERAL EXCAVATION

Under this heading the contractor is required to excavate the large areas for a cellar or basement below grade. This type of excavation is done with the aid of heavy mechanical equipment, such as the "steam shovel" or "bull dozer." From the plans the contractor can determine how far down (below grade) the excavation is to be carried. The contractor must make all required allowances beyond the foundation walls and retaining walls for the sloping of banks, forms for foundations, and waterproofing of the foundation walls.

To find the number of cubic yards (cu. yd.) of earth excavation in a basement or cellar, multiply the length by the width by the depth of the excavation. The result will be in cubic feet (cu. ft.). To find the cubic yards, divide the cubic feet by 27.

For example: Given the length of an excavation of 40'-0", the width 30'-0", and the depth 7'-0".

Then $40 \times 30 \times 7 = 8{,}400$ cu. ft.
or $8{,}400 \div 27 = 311$ cu. yd.

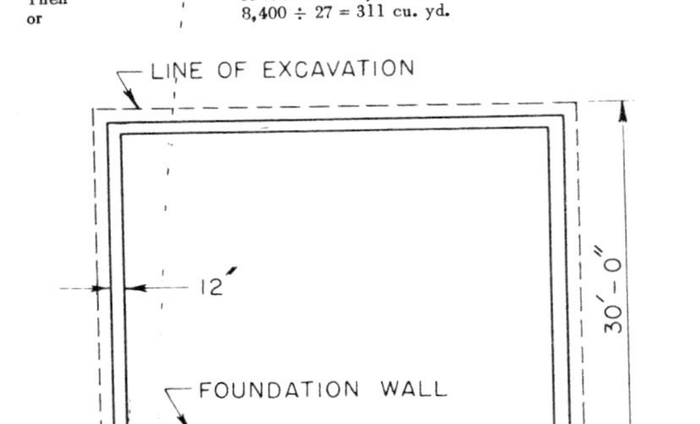

CELLAR EXCAVATION

Usually excavation is included in the masonry contract although sometimes it is let out under a special contract, which method will be followed in very large structures. The cost of excavating depends upon the soil, the job conditions, and the distance which the soil has to be moved from the job to a site where it can be dumped. Until such time as the structure is up, the excavated earth will be stored there, and then the process of back-filling will begin.

It is important for architect, engineer, contractor, and estimator alike to know the bearing power of the various kinds of soil.

Type of Soil	*Bearing Power, Tons per Sq. Ft.*
Rock	10
Hard clay or shale	8
Gravels or coarse sand	6
Clay or fine sand	3
Wet clay	2
Soft clay	1
Quicksand (sand and water)	½

Plans usually do not show the outline of the excavation or the ground level or the original grade of the site. Therefore it is necessary for the estimator to visit the site and determine the conditions that may affect the amount and size of the necessary excavations. Thereby he will be better enabled to visualize the actual amount of the excavation needed. Space must be allowed all around the building for proper working room for installing wood forms for all footings and foundation walls. In this connection the projection of the footings beyond the outside face of the walls has to be taken into consideration. Temporary guard rails, ramps, and other provisions which might be required by state or municipality for a safe execution of the work have to be included in the excavation estimate, and their erection, as well as the cement, has to be a part of the excavation estimate.

EXCAVATION TABLES

The tables on pages 16–29 serve as a quick means of finding the cubic yards of excavation without multiplying and dividing. The tables are for excavations ranging from sizes of 16' × 20' to 40' × 42', with depths ranging from 2' to 7' in increments of 3".

For example: To obtain the number of cubic yards of earth in a cellar 30' × 42', having a depth of 6'-9", turn to the table marked 6'-9" in the upper left-hand corner, and follow down the column of figures at the left-hand margin until the figure 30 is reached. Follow across the page (in row 30) until the figure under the column marked 42 is reached. The figure thus found represents the number of cubic yards of excavation for the cellar. The figure is 315 cu. yd. (to the nearest yard). If by actual multiplication and division a certain excavation were to be, say, 225.8 cu. yd., the table will show 226 cu. yd., or to the nearest full cubic yard.

If the same dimensions as above were used for an excavation, but with only half of the above depth, or 3'-4½" take half of the figure found for the 6'-9" depth. In other words,

A cellar with dimensions of 30' × 42' × 6'-9" = 315 cu. yd.
A cellar with dimensions of 30' × 42' × 3'-4½" = 158 cu. yd.

FIGURING EXCAVATION NOT CONTAINED IN TABLES

To figure sizes not given in the tables, take the following example: Assume an excavation depth of 6'-3". For a cellar 8' × 10', take ¼ of the result shown for a size of 16' × 20' (because 8' × 10' is ¼ of 16' × 20'.)

A cellar 16' × 20' × 6'-3" = 74 cu. yd. (from the table).
A cellar 8' × 10' × 6'-6" = 18.5 cu. yd. (¼ of 74).
An 8' × 10' cellar is ¼ of a 16' × 20' cellar.
A 10' × 10' cellar is ¼ of a 20' × 20' cellar..

For a cellar 13' × 20' take half of the results shown for a size of 26' × 20'.

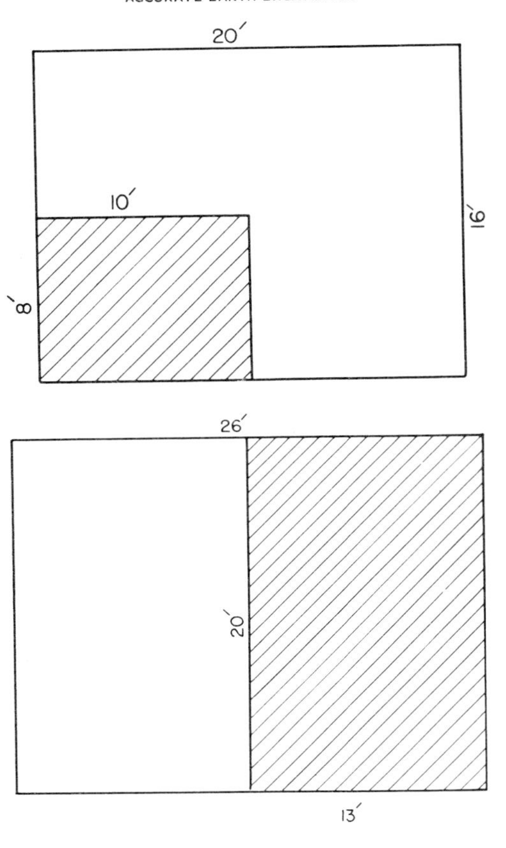

26' × 20' × 6'-3" = 60 cu. yd. (from the table). 13' × 20' × 6'-3" = 30 cu. yd. (½ of 60).

A 13' × 20' cellar is ½ of a 26' × 20' cellar.
A 12' × 27' cellar is ½ of a 24' × 27' cellar.

The rule therefore is: If both of the required dimensions equal one-half of a certain size given in the table, take one-fourth of the results shown.

If one of the required dimensions equals one-half of a certain size given in the table, take one-half of the result shown.

3'-9"

EXCAVATION TABLE IN CUBIC YARDS

For twice depth 7'-6"
take double results.

Dimensions of Excavation	20	21	22	23	24	25	26	27	28	29	30	31	32	33	34	35	36	37	38	39	40	41	42
16	44	47	49	51	53	56	58	60	62	64	67	69	71	73	76	78	80	82	84	87	89	91	93
17	47	50	52	54	57	59	61	64	66	68	71	73	76	78	80	83	85	87	90	92	94	97	99
18	50	52	55	57	60	62	65	68	70	72	75	77	80	82	85	87	90	92	95	97	100	102	105
19	53	55	58	61	63	66	69	71	74	76	79	82	84	87	90	92	95	98	100	103	105	108	111
20	56	58	61	64	67	69	72	75	78	81	83	86	89	92	94	97	100	103	106	108	111	114	117
21	58	61	64	67	70	73	76	79	82	85	87	90	93	96	99	102	105	108	111	114	117	120	122
22	61	64	67	70	73	76	79	83	86	89	92	95	98	101	104	107	110	113	116	119	122	125	128
23	64	67	70	73	77	80	83	86	89	93	96	99	102	105	109	112	115	118	121	125	128	131	134
24	67	70	73	77	80	83	87	90	93	97	100	103	107	110	113	117	120	123	127	130	133	137	140
25	69	73	76	80	83	87	90	94	97	101	104	108	111	115	118	121	125	128	132	135	139	142	146
26	72	76	79	83	87	90	94	98	101	105	108	112	116	119	123	126	130	134	139	141	144	148	152
27	75	79	82	86	90	94	97	101	105	109	112	116	120	124	127	131	135	139	142	146	150	154	157
28	78	82	86	89	93	97	101	105	109	113	117	121	124	128	132	136	140	144	148	152	156	159	163
29	81	85	89	93	97	101	105	109	113	117	121	125	129	133	137	141	145	140	153	157	161	165	169
30	83	87	92	96	100	104	108	113	117	121	125	129	133	137	142	146	150	154	158	162	167	171	175
31	86	90	95	99	103	108	112	116	121	125	129	133	138	142	146	151	155	159	164	168	172	176	181
32	89	93	98	102	107	111	115	120	124	129	133	138	142	147	151	155	160	164	169	173	178	182	187
33	92	96	101	105	110	114	119	124	128	133	137	142	147	151	156	160	165	170	179	179	183	188	192
34	94	99	104	109	113	118	123	128	132	137	142	146	151	156	160	165	170	175	185	184	189	194	198
35	97	102	107	111	117	121	126	131	136	141	146	151	155	160	165	170	175	180	190	190	194	199	204
36	100	105	110	115	120	125	130	135	140	145	150	155	160	165	170	175	180	185	190	195	200	205	210
37	103	108	113	118	123	128	134	139	144	149	154	159	164	169	175	180	185	190	195	200	206	211	216
38	106	111	116	121	127	132	137	143	148	153	158	164	169	174	179	185	190	195	201	206	211	216	222
39	108	114	119	124	130	135	141	146	152	157	162	168	173	179	184	189	195	200	206	211	217	222	227
40	111	117	122	128	133	139	144	150	156	161	167	172	178	183	189	194	200	206	211	217	222	228	223

4'-0"

For half depth 2'-0" take $\frac{1}{2}$ of results.

EXCAVATION TABLE IN CUBIC YARDS

Dimensions of Excavation	20	21	22	23	24	25	26	27	28	29	30	31	32	33	34	35	36	37	38	39	40	41	42
16	47	50	52	55	57	59	62	64	66	69	71	73	76	78	81	83	85	88	90	92	94	97	100
17	50	53	55	58	60	63	65	68	71	73	76	78	81	83	86	88	91	93	96	98	101	103	106
18	53	56	59	61	64	67	69	72	75	77	81	83	85	88	91	93	96	99	101	104	107	109	112
19	56	59	62	65	68	70	73	76	79	82	84	87	90	93	96	96	101	104	107	110	112	115	118
20	59	62	65	68	71	74	77	80	83	86	89	92	95	98	101	104	107	110	113	116	118	121	124
21	62	65	68	72	75	78	81	84	87	90	93	96	100	103	106	109	112	115	118	121	124	127	131
22	65	68	72	75	78	81	85	88	91	95	98	100	104	108	111	114	117	121	124	127	130	134	137
23	68	72	75	78	82	85	89	92	95	99	102	105	109	112	116	119	123	126	129	133	136	140	143
24	71	75	78	82	85	89	92	96	100	103	107	110	114	117	121	125	128	132	135	139	142	146	149
25	74	78	81	85	89	93	96	100	104	107	111	115	119	122	126	130	133	137	141	144	148	152	156
26	77	81	84	89	92	96	100	104	108	112	116	119	123	127	131	135	139	143	146	150	154	158	162
27	80	84	88	92	96	100	104	108	112	116	120	124	128	132	136	140	144	148	152	156	160	164	168
28	83	87	91	95	100	104	108	112	116	120	124	129	133	137	141	145	149	153	158	162	166	170	174
29	86	90	95	99	103	107	112	116	120	125	129	133	138	142	146	150	155	159	163	168	172	176	180
30	89	93	98	102	107	111	116	120	124	129	133	138	142	147	151	151	160	164	169	173	178	182	187
31	92	96	101	106	110	115	119	124	129	133	138	142	147	152	156	161	165	170	175	179	184	188	193
32	94	100	104	109	114	119	123	128	133	137	142	147	152	156	161	166	171	175	180	185	190	194	199
33	98	103	108	112	117	122	127	132	137	142	147	152	156	161	166	171	176	181	186	191	196	200	205
34	101	106	111	116	121	126	131	136	141	147	151	156	161	166	171	176	181	186	191	196	201	207	212
35	104	109	114	119	124	130	135	140	145	150	156	161	166	171	176	182	187	192	197	202	207	213	218
36	107	112	117	123	128	133	139	144	149	155	160	165	171	176	181	187	192	197	203	208	213	219	224
37	110	115	121	126	132	137	143	148	154	159	164	170	175	181	186	192	197	203	208	214	219	225	230
38	113	118	124	129	135	141	146	152	158	163	169	175	180	186	191	197	203	208	214	220	225	231	236
39	116	121	127	133	139	144	150	156	162	168	173	179	185	191	196	202	208	214	220	225	231	237	243
40	119	124	130	136	142	148	154	160	167	172	178	184	190	196	202	207	213	219	225	231	237	243	249

4'-3"

For half depth 2'-1$\frac{1}{2}$"
take $\frac{1}{2}$ of results.

EXCAVATION TABLE IN CUBIC YARDS

Dimensions of Excavation	20	21	22	23	24	25	26	27	28	29	30	31	32	33	34	35	36	37	38	39	40	41	42
16	50	53	55	58	60	63	66	68	71	73	76	78	81	83	86	88	91	93	96	98	101	103	106
17	54	56	59	62	64	67	70	72	75	76	80	83	86	88	91	94	96	99	102	104	107	110	112
18	57	59	62	65	68	71	74	77	79	82	85	88	91	94	96	99	102	105	108	110	113	116	119
19	60	63	66	69	72	75	78	81	84	87	90	93	96	99	102	105	108	111	114	117	120	123	126
20	63	66	69	72	76	79	82	85	88	91	94	98	101	104	107	110	113	116	120	123	126	129	132
21	66	69	73	76	79	83	86	89	93	96	99	103	106	109	112	116	119	122	126	129	132	135	139
22	69	73	76	80	83	86	90	94	97	100	104	107	111	114	118	121	125	128	132	135	138	142	145
23	72	76	80	83	87	90	94	98	101	105	109	112	116	119	123	127	130	134	138	141	145	148	152
24	76	80	83	87	91	94	98	102	106	110	113	117	121	125	128	132	136	140	144	147	151	155	159
25	77	83	87	90	94	98	102	106	110	114	118	122	126	130	134	138	142	146	150	153	157	161	165
26	82	86	90	94	98	102	106	111	115	119	123	127	131	135	139	143	147	151	156	160	164	168	172
27	85	89	93	98	102	106	110	115	119	123	127	132	136	140	144	149	153	157	162	166	170	174	179
28	88	93	97	101	106	110	115	119	123	128	132	137	141	145	150	154	159	163	167	172	176	181	185
29	91	96	100	105	110	114	119	123	128	132	137	142	146	151	155	160	164	169	173	178	183	187	192
30	94	99	103	109	113	118	123	128	132	137	142	146	151	156	161	165	170	175	179	184	189	194	198
31	98	102	107	112	117	122	127	132	137	141	146	151	156	161	166	171	176	181	185	190	195	200	205
32	101	106	111	116	121	126	131	136	141	146	151	156	161	166	171	176	181	186	191	196	201	207	212
33	104	109	114	119	125	130	135	140	145	151	156	161	166	171	177	182	187	192	197	203	208	213	218
34	107	112	118	123	128	134	139	145	150	155	160	166	171	177	182	187	193	198	203	209	214	220	225
35	110	116	121	127	132	138	143	149	154	160	165	171	176	182	187	193	198	204	209	215	220	226	231
36	113	119	125	130	136	142	147	153	159	164	170	176	181	187	193	198	204	210	215	221	227	232	238
37	117	122	128	134	140	145	151	157	163	169	175	181	186	192	198	204	210	215	221	227	233	239	245
38	120	126	132	138	143	149	155	162	167	173	179	185	191	197	203	209	215	221	227	233	239	245	251
39	123	129	135	141	147	153	160	166	172	178	184	190	196	203	209	215	221	227	233	239	246	252	258
40	126	132	139	145	151	157	164	170	176	183	189	195	201	208	214	220	227	233	239	246	252	258	264

4'-6"

EXCAVATION TABLE IN CUBIC YARDS

For half depth 2'-3"
take $\frac{1}{2}$ of results.

Dimensions of Excavation	20	21	22	23	24	25	26	27	28	29	30	31	32	33	34	35	36	37	38	39	40	41	42
16	53	56	59	61	64	67	69	72	75	77	80	83	85	88	91	93	96	99	101	104	107	109	112
17	57	60	62	65	68	71	74	77	79	82	85	88	91	94	96	99	102	105	108	111	113	116	119
18	60	63	66	69	72	75	78	81	84	87	90	93	96	99	102	105	108	111	114	117	120	123	126
19	63	67	70	73	76	79	82	86	89	92	95	98	101	105	108	111	114	117	120	124	127	130	133
20	67	70	73	77	80	83	87	90	93	97	100	103	107	110	113	117	120	123	127	130	133	137	140
21	70	74	77	80	84	87	91	95	98	101	105	108	112	116	119	123	126	130	133	137	140	144	147
22	73	77	81	84	88	92	95	99	103	106	110	114	117	121	125	128	132	136	139	143	147	150	154
23	77	81	84	88	92	96.	100	104	107	111	115	119	123	127	130	134	138	142	146	150	153	157	161
24	80	84	88	92	96	100	104	108	112	116	120	124	128	132	136	140	144	148	152	156	160	164	168
25	83	88	92	96	100	104	108	113	117	121	125	129	133	138	142	146	150	154	158	163	167	171	175
26	87	91	95	100	104	108	113	117	121	126	130	134	139	143	147	152	156	160	165	169	173	178	182
27	90	95	99	103	108	112	117	122	126	130	135	139	144	149	153	158	162	167	171	176	180	185	189
28	93	98	103	107	112	117	121	126	131	135	140	145	149	154	159	163	168	173	177	182	187	191	196
29	97	102	106	111	116	121	126	131	135	140	145	150	155	160	164	169	174	179	184	189	193	198	203
30	100	105	110	115	120	125	130	135	140	145	150	155	160	165	170	175	180	185	190	195	200	205	210
31	103	109	114	119	124	129	134	140	145	150	155	160	165	171	176	181	186	191	196	202	207	212	217
32	107	112	117	123	128	133	139	144	149	155	160	165	171	176	181	187	192	197	203	208	213	219	224
33	110	116	121	126	132	137	143	149	154	159	165	170	176	182	187	193	198	204	209	215	220	226	231
34	113	119	125	130	136	142	147	153	159	164	170	176	181	187	193	198	204	210	215	221	227	232	238
35	117	123	128	134	140	146	152	158	163	169	175	181	187	193	198	204	210	216	222	228	233	239	245
36	120	126	132	138	144	150	156	162	168	174	180	186	192	198	204	210	216	222	228	234	240	246	252
37	123	130	136	142	148	154	160	167	173	179	185	191	197	204	210	216	222	228	234	241	247	253	259
38	127	133	139	146	152	158	165	171	177	184	190	196	203	209	215	222	228	234	241	247	253	260	266
39	130	137	143	149	156	162	169	176	182	188	195	202	208	215	221	227	234	241	247	254	260	266	273
40	133	140	147	153	160	167	173	180	187	193	200	207	213	220	227	233	240	247	253	260	267	273	280

4'-9"

EXCAVATION TABLE IN CUBIC YARDS

For half depth 2'-4$\frac{1}{2}$"
take $\frac{1}{2}$ of results.

Dimensions of Excavation	20	21	22	23	24	25	26	27	28	29	30	31	32	33	34	35	36	37	38	39	40	41	42
16	56	59	62	65	68	70	73	76	79	82	84	87	90	93	96	99	101	104	107	110	113	115	118
17	60	63	66	69	72	75	78	81	84	87	90	93	96	99	102	105	108	111	114	117	120	123	126
18	63	66	70	73	76	79	82	86	89	92	95	98	101	105	108	111	114	117	120	124	127	130	133
19	67	70	74	77	80	84	87	90	94	97	100	104	107	110	114	117	120	124	127	130	134	137	140
20	70	74	77	81	84	88	91	95	98	102	106	109	113	116	120	123	127	130	134	137	141	144	148
21	74	78	81	85	89	92	96	100	103	107	111	115	118	122	126	129	133	137	140	144	148	151	155
22	77	81	85	90	93	97	101	105	108	112	116	120	124	128	132	135	139	143	147	151	155	159	163
23	81	85	90	93	97	101	105	109	113	117	121	125	129	134	138	142	146	150	154	158	162	166	170
24	84	89	93	97	101	106	110	114	118	122	127	131	135	139	144	148	152	156	160	165	169	173	177
25	88	92	97	101	106	110	114	119	123	128	132	136	141	145	150	154	158	163	167	172	176	180	185
26	92	96	101	105	110	114	119	124	128	133	137	142	146	151	156	160	165	169	174	178	183	188	192
27	95	100	104	109	114	119	123	128	133	138	143	147	152	157	161	166	171	176	180	185	190	195	200
28	99	103	108	113	118	123	128	133	138	143	148	153	158	163	167	172	177	182	187	192	197	202	207
29	102	107	112	117	122	128	133	138	143	148	153	158	163	168	173	179	184	189	194	199	204	209	214
30	106	111	116	121	127	132	137	143	148	153	158	164	169	174	179	185	190	195	200	206	211	216	222
31	109	114	120	125	131	136	142	147	153	158	164	169	174	180	185	191	196	202	207	213	218	224	229
32	113	118	124	129	135	141	146	152	158	163	169	175	180	186	191	197	203	208	214	220	225	231	236
33	116	122	128	133	139	145	151	157	162	168	174	180	186	192	197	203	209	215	220	226	232	238	244
34	120	126	132	138	143	150	155	162	167	173	179	185	191	197	203	209	215	221	227	233	239	245	251
35	123	129	135	142	148	154	160	166	172	179	185	191	197	203	209	216	222	228	234	240	246	252	259
36	127	133	139	146	152	158	165	171	177	184	190	196	203	209	215	222	228	234	241	247	253	260	266
37	130	137	143	150	156	163	169	176	182	189	195	202	208	215	221	228	234	241	247	254	260	267	273
38	134	140	147	154	160	167	174	181	187	194	200	207	214	221	227	234	241	247	254	261	267	274	281
39	137	144	151	158	165	172	178	185	192	199	206	213	220	226	233	240	247	254	261	268	274	281	288
40	141	148	155	162	169	176	183	190	197	204	211	218	225	232	239	246	253	260	267	274	281	288	296

5'-0"

For half depth 2'-6" take $\frac{1}{2}$ of results.

Dimensions of Excavation	20	21	22	23	24	25	26	27	28	29	30	31	32	33	34	35	36	37	38	39	40	41	42
16	59	62	65	68	71	74	77	80	83	86	89	92	95	98	101	104	107	110	113	116	119	122	124
17	63	66	69	72	76	79	82	85	88	91	94	98	101	104	107	110	113	116	120	123	126	129	132
18	67	70	73	77	80	83	87	90	93	97	100	103	107	110	113	117	120	123	127	130	133	137	140
19	70	74	77	81	84	88	91	95	98	102	106	109	113	116	120	123	127	130	134	137	141	144	148
20	74	78	81	85	89	93	96	100	104	107	111	115	118	122	126	130	133	137	141	144	148	152	156
21	78	82	86	89	93	97	101	105	109	113	117	121	124	128	132	136	140	144	148	152	156	159	163
22	82	86	90	94	98	102	106	110	114	118	122	126	130	134	139	143	147	151	155	159	163	167	171
23	85	89	92	98	102	107	111	115	119	123	128	132	136	141	145	149	153	158	162	166	170	175	179
24	89	93	98	102	107	111	115	120	124	129	133	138	142	147	151	156	160	164	169	173	178	182	187
25	93	97	102	106	111	116	120	125	130	134	139	143	148	153	157	162	167	171	176	181	185	190	194
26	96	101	106	111	116	120	125	130	135	140	144	149	154	159	164	169	173	178	183	188	193	197	202
27	100	105	110	115	120	125	130	135	140	145	150	155	160	165	170	175	180	185	190	195	200	205	210
28	104	109	114	119	124	130	135	140	145	150	156	161	166	171	176	181	187	192	197	202	207	213	218
29	107	113	118	123	129	134	140	145	150	156	161	166	172	177	183	188	193	199	204	209	215	220	226
30	111	117	122	128	133	139	144	150	156	161	167	172	178	183	189	194	200	206	211	217	222	228	233
31	115	121	126	132	138	144	149	155	161	166	172	178	184	189	195	201	207	212	218	224	230	235	240
32	119	124	130	136	142	148	154	160	166	172	178	184	190	196	202	207	213	219	225	231	237	243	247
33	122	128	134	141	147	153	159	165	171	177	183	189	195	202	208	214	220	226	232	238	244	251	257
34	126	132	133	145	151	157	164	170	176	183	189	195	201	208	214	220	227	233	239	246	252	258	264
35	130	136	143	149	156	162	168	175	182	188	194	201	207	214	220	227	233	240	246	253	259	266	272
36	133	140	147	153	160	167	173	180	187	193	200	207	213	220	227	233	240	247	253	260	267	273	280
37	137	144	151	158	165	171	178	185	192	199	205	212	219	226	233	240	247	253	260	267	274	281	288
38	141	148	155	162	170	176	183	190	197	204	211	218	225	232	239	246	253	260	267	274	281	288	296
39	144	152	159	166	173	181	188	195	202	209	217	224	231	238	246	253	260	267	275	282	289	296	303
40	148	156	163	170	178	185	193	200	207	215	222	230	237	244	252	259	267	274	282	289	296	304	311

EXCAVATION TABLE IN CUBIC YARDS

For half depth 2'-7$\frac{1}{2}$"
take $\frac{1}{2}$ of results.

Dimensions of Excavation	20	21	22	23	24	25	26	27	28	29	30	31	32	33	34	35	36	37	38	39	40	41	42
16	62	65	68	72	75	78	81	84	87	90	93	96	100	103	106	109	112	115	118	121	124	126	131
17	66	69	73	76	79	83	86	89	93	96	99	102	106	109	112	116	119	122	126	129	132	135	139
18	70	73	77	81	84	88	91	95	98	101	105	108	112	115	119	123	126	130	133	136	140	143	147
19	74	78	81	85	89	92	96	100	103	107	111	114	118	122	126	129	133	137	140	144	148	151	155
20	78	82	86	89	93	97	101	105	109	113	117	121	124	128	132	135	140	144	148	152	156	159	163
21	82	86	90	94	98	102	106	110	114	118	122	127	131	135	139	143	147	151	155	159	163	167	171
22	85	90	94	98	103	107	111	116	120	124	128	133	137	141	146	150	154	158	163	167	171	175	181
23	89	94	98	103	107	112	116	121	125	130	134	139	143	148	152	157	161	166	170	174	179	183	188
24	93	98	103	107	112	117	121	126	131	135	140	145	149	154	159	163	168	173	177	182	187	191	196
25	97	102	107	112	117	122	126	131	136	141	146	151	155	161	165	170	175	180	185	190	194	199	204
26	101	106	111	116	121	126	131	137	142	147	152	157	162	167	172	177	182	187	192	197	202	207	212
27	105	110	115	121	126	131	136	142	147	152	157	163	168	173	179	184	189	194	199	205	210	215	220
28	109	114	120	125	131	136	142	147	152	158	163	169	174	180	185	191	196	202	207	212	218	223	229
29	113	118	124	130	135	141	147	152	158	164	169	175	180	186	192	197	203	209	214	220	226	231	237
30	117	122	128	134	140	146	152	158	162	169	175	181	187	193	198	204	210	216	222	228	233	239	245
31	121	127	133	139	145	151	157	163	169	175	181	187	193	199	205	211	217	223	229	235	241	247	253
32	124	131	137	143	149	156	162	168	174	180	187	193	199	205	212	218	224	230	236	243	249	255	261
33	128	135	141	148	154	160	167	173	180	186	192	199	205	212	218	225	231	237	244	250	257	263	269
34	132	139	145	152	159	165	172	179	185	192	198	205	211	218	225	231	238	245	251	258	264	271	278
35	136	143	150	157	163	170	177	184	190	197	204	211	218	225	232	238	245	252	259	266	272	279	286
36	140	147	154	161	168	175	182	189	196	203	210	217	224	231	238	245	252	259	266	273	280	287	294
37	144	151	158	165	173	180	187	194	201	209	216	223	230	238	245	252	259	266	273	281	288	295	302
38	148	155	163	170	177	185	192	200	207	214	222	229	236	244	251	259	266	273	281	288	296	303	310
39	152	159	167	174	182	190	197	205	212	220	227	235	243	250	258	266	273	281	288	296	303	311	319
40	156	163	171	179	187	195	202	210	218	226	233	241	249	257	265	272	280	288	296	303	311	319	327

5'-6"

EXCAVATION TABLE IN CUBIC YARDS

For half depth 2'-9" take $\frac{1}{2}$ of results.

Dimensions of Excavation	20	21	22	23	24	25	26	27	28	29	30	31	32	33	34	35	36	37	38	39	40	41	42
16	65	69	72	75	78	82	85	88	91	95	98	101	104	108	111	114	117	121	124	127	130	134	137
17	69	73	76	80	83	87	90	94	97	100	104	107	111	114	118	121	125	128	132	136	139	142	145
18	73	77	81	84	88	92	95	99	103	106	110	114	117	121	125	128	133	136	139	143	147	150	154
19	77	81	85	89	93	97	101	105	108	112	116	120	124	128	132	135	139	143	147	151	155	159	163
20	81	86	90	94	98	102	106	110	114	118	122	126	130	134	139	143	147	151	155	159	163	167	171
21	86	90	94	98	103	107	111	116	120	124	128	133	137	141	145	150	154	158	163	167	171	175	180
22	90	94	99	103	108	112	117	121	126	130	134	139	143	148	152	157	161	166	170	175	179	184	188
23	94	98	103	108	112	117	122	127	131	136	141	145	150	155	159	164	169	173	178	183	187	192	197
24	98	103	108	112	117	122	127	132	137	142	147	152	156	161	166	171	176	181	186	191	196	201	205
25	102	107	112	117	122	127	132	138	143	148	153	158	163	168	173	178	183	188	194	199	204	209	214
26	106	111	117	122	127	132	138	143	148	154	159	164	170	175	180	185	191	196	201	207	212	217	222
27	110	115	121	126	132	137	143	149	154	160	165	171	176	182	187	193	198	204	209	214	220	226	231
28	114	120	125	131	137	142	148	154	160	165	171	177	183	188	194	200	205	211	217	222	228	234	240
29	118	124	130	136	142	148	154	160	165	171	177	183	189	195	201	207	213	219	225	233	236	242	248
30	122	128	134	141	147	153	159	165	171	177	184	189	196	202	208	214	220	226	232	238	245	251	257
31	126	133	139	145	152	158	164	171	177	183	189	196	202	208	215	221	227	234	240	246	253	258	265
32	130	137	143	150	156	163	169	176	183	189	196	202	209	215	222	228	235	241	248	254	261	267	274
33	134	141	148	155	161	168	175	182	188	195	202	208	215	222	229	235	242	249	255	262	269	276	282
34	138	145	152	159	166	173	180	187	194	201	208	215	222	229	236	242	249	256	263	270	277	284	291
35	143	150	157	164	171	178	185	193	200	207	214	221	228	235	242	250	257	264	271	278	285	292	299
36	147	154	161	169	176	183	191	198	205	213	220	227	235	242	249	257	264	271	279	286	293	301	308
37	151	158	166	173	181	188	196	204	211	219	226	234	241	249	256	264	271	279	287	294	301	309	316
38	155	163	170	178	186	193	201	209	217	225	232	240	248	255	263	271	279	286	294	302	310	317	326
39	159	167	175	183	191	199	206	215	222	230	238	246	254	262	270	278	286	294	302	310	318	326	334
40	163	171	179	187	196	204	212	220	228	236	244	253	261	269	277	285	293	302	310	318	326	334	342

EXCAVATION TABLE IN CUBIC YARDS

For half depth 2'-10$\frac{1}{2}$"
take $\frac{1}{2}$ of results.

Dimensions of Excavation	20	21	22	23	24	25	26	27	28	29	30	31	32	33	34	35	36	37	38	39	40	41	42
16	68	72	75	78	82	85	89	92	95	99	102	106	109	112	116	119	123	126	129	133	136	140	143
17	72	76	80	83	87	90	94	98	101	105	109	112	116	119	123	127	130	134	137	141	145	148	152
18	77	81	84	88	92	96	96	104	107	111	115	119	123	126	130	134	138	142	146	149	153	157	161
19	81	85	89	93	97	101	105	109	113	117	121	125	129	133	138	142	146	150	154	158	162	166	170
20	85	89	94	98	102	106	111	115	119	124	128	132	136	140	145	149	153	158	162	166	170	175	179
21	89	94	98	103	107	112	116	121	125	130	134	139	143	147	152	156	161	165	170	174	179	183	188
22	94	98	103	108	112	117	122	127	131	136	141	145	150	154	159	164	169	173	178	183	187	192	197
23	98	103	108	113	118	122	127	132	137	142	147	152	157	162	166	171	176	181	186	191	196	201	206
24	102	107	112	118	123	128	133	138	143	148	153	158	164	169	174	179	184	189	195	199	204	209	215
25	106	112	117	122	128	133	138	144	149	154	160	165	170	176	181	186	192	197	202	208	213	218	224
26	111	116	122	127	133	138	144	150	155	161	166	172	177	183	188	194	199	205	210	216	221	227	233
27	115	121	126	132	138	144	149	155	161	167	172	178	184	190	195	201	207	213	218	224	230	236	241
28	119	125	131	137	143	149	155	161	167	173	179	185	191	197	203	209	215	221	226	232	238	244	250
29	123	130	136	142	148	154	161	167	173	179	185	191	198	204	210	216	222	228	235	241	247	253	259
30	128	134	141	147	153	160	166	173	179	185	192	198	204	211	217	222	230	236	243	249	255	262	268
31	132	139	145	152	159	165	172	178	185	192	198	205	211	218	224	231	238	244	251	257	264	271	277
32	136	143	150	157	164	170	177	184	191	198	204	211	218	225	232	238	245	252	259	266	274	279	286
33	141	148	155	162	169	176	183	190	197	204	211	218	225	232	239	246	253	260	267	274	281	288	295
34	145	152	159	167	174	181	188	196	203	210	217	224	232	239	246	253	260	268	275	282	290	297	304
35	149	157	164	171	179	186	194	201	209	216	224	231	238	246	253	261	268	276	283	291	298	305	313
36	153	161	169	176	184	192	199	207	215	222	230	238	245	253	261	268	276	284	291	299	307	314	322
37	158	165	173	181	189	197	205	213	221	229	236	244	252	260	268	276	283	291	299	307	315	323	331
38	162	170	178	186	194	202	210	219	227	235	243	251	259	267	275	283	291	299	307	315	324	332	340
39	166	174	183	191	199	208	216	224	232	241	249	257	266	274	282	291	299	307	315	324	332	340	349
40	170	179	187	196	204	213	221	230	238	247	256	264	273	281	290	298	307	315	324	332	341	349	358

6'-0"

For half depth 3'-0"
take $\frac{1}{2}$ of results.

EXCAVATION TABLE IN CUBIC YARDS

Dimensions of Excavation	20	21	22	23	24	25	26	27	28	29	30	31	32	33	34	35	36	37	38	39	40	41	42
16	71	75	78	82	85	89	93	96	100	103	107	110	114	117	121	124	128	132	135	139	142	146	149
17	76	79	83	87	91	94	98	102	106	110	113	117	121	125	128	132	136	140	144	147	151	155	159
18	80	84	88	92	96	100	104	108	112	116	120	124	128	132	136	140	144	148	152	156	160	164	168
19	84	89	93	97	101	105	110	114	118	122	127	131	135	139	144	148	152	156	160	165	169	173	177
20	89	93	98	102	107	111	116	120	124	129	133	138	142	147	151	156	160	164	169	173	178	182	187
21	93	98	103	107	112	117	121	126	131	135	140	145	149	154	159	163	168	173	173	182	187	191	196
22	98	103	108	112	117	122	127	132	137	142	147	152	156	161	166	171	176	181	186	191	196	200	205
23	102	107	112	118	123	128	133	138	143	148	153	158	164	169	174	179	184	189	194	199	204	210	215
24	107	112	117	123	128	133	139	144	149	155	160	165	171	176	181	187	192	197	203	208	213	219	224
25	111	117	122	128	133	139	144	150	156	161	167	172	178	183	189	194	200	206	211	217	222	228	233
26	116	121	127	133	139	144	150	156	162	168	173	179	185	191	196	202	208	214	220	225	231	237	243
27	120	126	132	138	144	150	156	162	168	174	180	186	192	198	204	210	216	222	228	234	240	246	252
28	124	131	137	143	149	156	162	168	174	180	187	193	199	205	212	218	224	230	236	243	249	255	261
29	129	135	142	148	155	161	168	174	180	187	193	200	206	213	219	226	232	238	245	251	258	264	271
30	133	140	147	153	160	167	173	180	187	193	200	207	213	220	227	233	240	247	253	260	267	273	280
31	138	145	152	158	165	172	179	186	193	200	207	214	220	227	234	241	248	255	262	269	276	282	289
32	142	149	156	164	171	178	185	192	199	206	213	220	228	235	242	249	256	263	270	277	284	292	299
33	147	154	161	169	176	183	191	198	205	213	220	227	235	242	249	257	264	271	279	286	293	301	308
34	151	159	166	174	181	189	196	204	212	219	227	234	242	249	257	264	272	280	287	295	302	310	317
35	156	164	171	179	187	194	202	210	218	225	233	241	249	257	264	272	280	288	294	303	311	319	327
36	160	168	176	184	192	200	208	216	224	232	240	248	256	264	272	280	288	296	304	312	320	326	336
37	164	173	181	189	197	205	214	222	230	238	247	255	263	271	279	288	296	304	312	321	329	337	345
38	169	177	186	194	203	211	220	228	236	245	253	262	270	279	287	296	304	312	321	329	338	346	355
39	173	182	191	199	208	217	225	234	243	251	260	269	277	286	295	303	312	321	329	338	347	355	364
40	178	187	196	204	213	222	231	240	249	258	267	276	284	293	302	311	320	329	338	347	356	364	373

6'-3"

For half depth 3'-3"
take $\frac{1}{2}$ of results.

EXCAVATION TABLE IN CUBIC YARDS

Dimensions of Excavation	20	21	22	23	24	25	26	27	28	29	30	31	32	33	34	35	36	37	38	39	40	41	42
16	74	78	82	85	89	93	96	100	104	107	111	115	119	122	126	130	133	137	141	144	148	152	156
17	79	83	87	91	94	98	102	106	110	114	118	122	126	130	134	138	142	146	150	153	157	161	165
18	83	88	92	96	100	104	108	113	117	121	125	129	133	137	142	146	150	154	158	162	167	171	175
19	88	92	97	101	106	110	114	119	123	128	132	136	141	145	150	154	158	163	167	171	176	180	185
20	93	97	102	106	111	116	120	125	130	134	139	144	148	153	157	162	167	171	176	180	185	190	194
21	97	102	107	112	117	122	126	131	136	141	146	151	156	160	165	170	175	180	185	190	194	199	204
22	102	107	112	117	122	127	132	138	142	148	153	158	163	168	173	178	183	188	194	199	204	209	214
23	107	112	117	122	128	133	138	144	149	154	160	165	170	176	181	186	192	197	202	208	213	218	224
24	111	117	122	128	133	139	144	150	155	161	167	172	178	183	189	194	200	206	211	217	222	228	233
25	116	122	127	133	139	145	150	156	162	168	174	179	185	191	197	203	208	214	220	226	231	237	243
26	120	126	132	138	144	151	157	163	168	175	181	187	193	199	205	211	217	223	229	235	241	247	253
27	125	131	137	144	150	156	163	169	175	181	187	194	200	206	212	219	225	231	238	244	250	256	262
28	130	136	143	149	156	162	169	175	181	188	194	201	207	214	220	227	233	240	246	253	259	266	272
29	134	141	148	154	162	168	175	181	188	195	201	208	215	222	228	235	242	248	255	262	268	275	282
30	139	146	153	160	167	174	181	188	194	201	208	215	222	229	236	243	250	257	264	271	278	285	292
31	144	151	158	165	172	179	187	194	201	208	215	223	230	237	244	251	258	266	273	280	287	294	301
32	148	156	163	170	178	185	193	200	207	215	222	230	237	244	252	259	267	274	282	289	296	304	311
33	153	160	168	176	183	191	199	206	214	221	229	237	244	252	260	267	275	283	290	298	306	313	321
34	157	165	173	181	189	197	205	213	220	228	236	244	252	260	268	275	283	291	299	307	315	323	330
35	162	170	178	186	194	203	211	219	227	235	243	251	259	267	275	224	292	300	308	316	324	332	340
36	167	175	183	192	200	208	217	225	233	242	250	258	267	275	283	292	300	308	317	325	333	342	350
37	171	180	188	197	205	214	223	231	240	248	257	266	274	283	291	300	308	317	326	334	343	351	360
38	176	185	193	202	211	220	229	238	246	255	264	273	282	290	299	308	317	326	334	343	352	361	369
39	181	190	199	208	217	226	235	244	253	262	271	280	289	298	307	316	325	334	343	352	361	370	379
40	185	194	204	213	222	231	241	250	259	269	278	287	296	306	315	324	333	343	352	361	370	380	389

6'-6"

For half depth 3'-3" take $\frac{1}{2}$ of results.

EXCAVATION TABLE IN CUBIC YARDS

Dimensions of Excavation	20	21	22	23	24	25	26	27	28	29	30	31	32	33	34	35	36	37	38	39	40	41	42
16	77	81	85	89	92	96	100	104	108	112	116	119	123	127	131	135	139	143	146	150	154	158	162
17	82	86	90	94	98	102	106	111	115	119	123	127	131	135	139	143	147	151	155	160	164	168	172
18	87	91	95	100	104	108	113	117	121	126	130	134	139	143	142	152	156	160	165	169	173	178	182
19	91	96	101	105	110	114	119	124	128	132	137	142	146	151	155	160	165	169	174	178	183	187	192
20	96	101	106	111	116	120	125	130	135	140	144	149	154	159	164	168	173	178	183	188	193	197	202
21	101	106	111	116	121	126	131	137	142	147	152	157	162	167	172	177	182	187	192	197	202	207	212
22	106	111	117	122	127	132	138	143	148	153	159	164	169	175	180	188	191	196	201	206	212	217	222
23	111	116	122	127	133	138	144	150	155	160	166	172	177	183	188	194	199	205	210	216	221	227	232
24	116	121	127	133	139	144	150	156	162	167	173	179	185	191	196	202	208	214	220	225	236	237	243
25	120	126	132	138	144	150	156	163	168	174	180	187	193	199	205	211	217	223	229	235	241	247	253
26	125	131	138	144	150	156	163	169	175	181	188	194	200	206	213	219	225	232	238	244	250	257	263
27	130	136	143	149	156	162	169	176	182	189	195	202	208	214	221	227	234	241	247	253	260	266	273
28	135	142	148	155	162	168	175	182	189	195	202	209	216	222	229	236	243	249	251	263	270	276	283
29	140	147	154	161	168	174	181	189	195	202	209	216	223	230	237	244	251	258	265	272	279	286	293
30	143	152	159	166	173	180	188	195	202	209	217	224	231	238	246	253	260	267	274	282	289	296	303
31	149	157	164	172	179	187	194	202	209	216	224	231	239	246	254	261	269	276	284	290	298	306	313
32	154	162	170	177	185	193	200	208	216	223	231	239	246	254	262	270	277	285	293	300	308	316	323
33	159	167	175	183	191	199	207	215	222	230	238	246	254	262	270	278	286	294	302	310	318	326	334
34	164	172	180	188	196	205	213	221	229	237	245	254	262	270	278	286	295	303	311	319	327	335	344
35	168	177	185	194	202	211	219	228	236	244	253	261	270	278	286	295	303	312	320	329	337	345	354
36	173	182	191	199	208	217	225	234	243	251	260	269	277	286	295	303	312	321	329	338	347	355	364
37	178	187	196	205	214	223	232	241	249	258	267	276	285	294	303	312	321	330	338	347	356	364	374
38	183	192	201	210	220	229	238	247	256	265	274	284	293	302	311	320	329	339	348	357	366	375	384
39	188	197	207	216	225	235	244	254	263	272	282	291	300	310	319	328	338	347	357	366	375	385	394
40	193	202	212	221	231	241	250	260	270	279	289	299	308	318	327	337	347	356	366	376	385	395	404

6'-9"

EXCAVATION TABLE IN CUBIC YARDS

For half depth 3'-4$\frac{1}{2}$"
take $\frac{1}{2}$ of results.

Dimensions of Excavation	20	21	22	23	24	25	26	27	28	29	30	31	32	33	34	35	36	37	38	39	40	41	42
16	80	84	88	92	96	100	104	108	112	116	120	124	128	132	136	140	144	148	152	156	160	164	168
17	85	89	94	98	102	106	111	115	119	123	128	132	136	140	145	149	153	157	162	166	170	174	179
18	90	95	99	104	108	113	117	122	126	131	135	140	144	149	153	158	162	167	171	176	180	185	189
19	95	100	105	109	114	119	124	128	133	138	143	147	152	157	162	166	171	176	181	185	190	195	200
20	100	105	110	115	120	125	130	135	140	145	150	155	160	165	170	175	180	185	190	195	200	205	210
21	105	110	116	121	126	131	137	142	147	152	158	163	168	173	179	184	189	194	200	205	210	215	221
22	110	116	121	127	132	138	143	148	154	160	165	171	176	182	187	193	198	204	209	215	220	226	231
23	115	121	127	132	138	144	150	155	161	167	173	178	184	190	196	201	207	213	219	224	230	236	242
24	120	126	132	138	144	150	156	162	168	174	180	186	194	198	204	210	216	222	228	234	240	246	252
25	125	131	138	144	150	156	163	169	175	181	188	194	200	206	213	219	225	231	238	244	250	256	263
26	130	137	143	150	156	163	169	176	182	189	195	202	208	215	221	228	234	241	247	254	260	267	273
27	135	142	149	155	162	169	176	182	189	196	203	209	216	223	230	236	243	250	257	263	270	277	284
28	140	147	154	161	168	175	182	189	196	203	210	217	224	231	238	245	252	259	266	273	280	287	294
29	145	152	160	167	174	181	189	196	203	210	218	225	232	239	247	254	261	268	276	283	290	297	305
30	150	158	165	173	180	188	195	203	210	218	225	233	240	248	255	263	270	278	285	293	300	308	315
31	155	163	171	178	186	194	202	209	217	225	233	240	248	256	264	271	279	287	295	302	310	318	326
32	160	168	176	184	192	200	208	216	224	232	240	248	256	264	272	280	288	296	304	312	320	328	336
33	165	173	182	190	198	206	215	223	231	239	248	256	264	272	281	289	297	305	314	322	330	338	347
34	170	179	187	196	204	213	221	230	238	247	255	264	272	281	289	298	306	315	323	332	340	349	357
35	175	184	193	201	210	219	228	236	245	254	263	271	280	289	298	306	315	324	333	341	350	359	368
36	180	189	198	207	216	225	234	243	252	261	270	279	288	297	306	315	324	333	342	351	360	369	378
37	185	194	204	213	222	231	241	250	259	268	278	287	296	305	315	324	333	342	352	361	370	379	389
38	190	200	209	219	228	238	247	257	266	276	285	295	304	314	323	333	342	352	361	371	380	390	399
39	195	205	215	224	234	244	254	263	273	283	293	302	312	322	332	341	351	361	371	380	390	400	410
40	200	210	220	230	240	250	260	270	280	290	300	310	320	330	340	350	360	370	380	390	400	410	420

7'-0"

EXCAVATION TABLE IN CUBIC YARDS

For half depth 3'-6" take $\frac{1}{2}$ of results.

Dimensions of Excavation	20	21	22	23	24	25	26	27	28	29	30	31	32	33	34	35	36	37	38	39	40	41	42
16	83	87	91	95	100	104	108	112	116	120	124	129	133	137	141	145	149	154	158	162	166	170	174
17	88	93	97	101	106	110	115	119	123	128	132	137	141	145	150	154	159	163	167	172	176	181	185
18	93	98	103	107	112	117	121	126	131	135	140	145	149	154	159	163	168	173	177	182	187	191	196
19	98	103	108	113	118	123	128	133	138	143	148	153	158	163	167	172	177	182	187	192	197	202	207
20	104	109	114	119	124	130	135	140	145	150	156	161	166	171	176	181	187	192	197	202	207	213	218
21	109	114	120	125	131	136	142	147	152	158	163	169	174	180	186	191	196	201	207	212	218	223	229
22	114	120	125	131	137	143	148	154	160	165	171	177	182	188	194	200	205	211	217	222	228	234	240
23	119	125	131	137	143	149	155	161	167	173	179	185	191	197	203	209	215	221	227	233	238	244	250
24	124	131	137	143	149	156	162	167	174	180	187	193	199	205	211	218	224	230	236	243	249	255	261
25	130	136	143	149	156	162	168	175	181	188	194	201	207	214	220	227	233	240	246	253	259	266	272
26	135	142	148	155	162	169	175	182	189	196	202	209	216	222	229	236	243	249	256	263	270	276	283
27	140	147	154	161	168	175	182	189	196	203	210	217	224	231	238	245	252	259	266	273	280	287	294
28	145	152	160	167	174	181	189	196	203	211	218	225	232	240	247	254	261	269	276	283	290	298	305
29	150	158	165	173	180	188	195	203	210	218	226	233	240	248	256	263	271	278	286	293	301	308	316
30	156	163	171	179	187	194	202	210	218	226	233	241	249	257	264	272	280	288	296	303	311	319	327
31	161	169	171	185	193	201	209	217	225	233	241	249	257	265	273	281	289	297	305	313	321	329	338
32	166	174	182	191	199	107	216	224	232	241	249	257	265	274	282	290	298	307	315	323	332	340	348
33	171	180	188	197	205	214	222	231	240	248	257	265	274	282	291	299	308	317	325	334	242	351	359
34	176	185	194	203	212	220	229	238	247	256	264	273	282	291	300	308	317	326	335	344	353	361	370
35	182	190	200	209	218	227	236	245	254	263	272	281	290	299	308	318	326	336	345	354	363	372	381
36	187	196	205	215	224	233	243	252	261	271	280	289	299	308	317	327	336	345	355	364	373	383	392
37	192	201	211	221	230	240	249	259	269	278	287	297	307	316	326	336	345	355	364	374	384	393	403
38	197	207	217	227	236	246	256	266	276	286	296	305	315	325	345	345	354	364	374	384	394	404	414
39	202	212	222	233	243	253	263	273	283	293	303	313	323	334	344	354	364	374	384	394	404	414	425
40	207	218	228	239	249	259	270	280	290	301	311	321	332	342	353	363	373	384	394	404	415	425	436

FIGURING LARGER EXCAVATIONS NOT CONTAINED IN TABLES

To illustrate how to figure larger sizes than listed in the tables, take the following examples. Assume an excavation depth of 6'-3".

1. For a cellar 22' × 100', multiply the result shown in the table for a size 22' × 25' × 4. The 22' × 100' cellar is four times the size of the 22' × 25' cellar. Therefore,

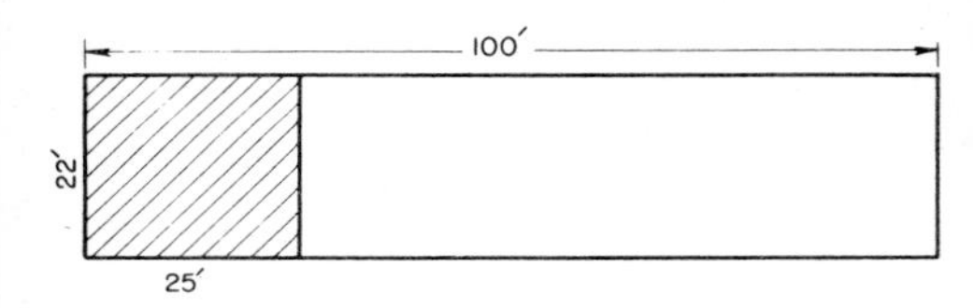

A cellar 22' × 25' × 6'-3" = 127 cu. yd. (from table).
A cellar 22' × 100' × 6'-3" = 508 cu. yd. (4 × 127).
A 22' × 100' cellar is 4 times a 22' × 25' cellar.

2. For a cellar 30' × 47', add together the result shown in the table for a 23' × 30' and a 24' × 30', as 23' plus 24' equals 47'. Therefore,

23' × 30' × 6'-3" = 160 cu. yd.
24' × 30' × 6'-3" = 167 cu. yd.

Total = 327 cu. yd. for a 30' × 47' size

3. For a cellar 44' × 120', take the result shown in the table at 22' × 40', and multiply by 6, as 44 is double 22' and 120' is three times the length of 40. Therefore,

22' × 40' × 6'-3" = 204 cu. yd. (from table).
44' × 120' × 6'-3" = 1,224 cu. yd. (6 × 204).

4. For a cellar 30' × 49', double the result shown for a cellar 24' × 30'.

From these examples it will be seen that almost any size may be readily figured.

FIGURING EXCAVATIONS WITH DEPTH NOT CONTAINED IN TABLES

To figure odd inches in depth, which are not shown in the tables, such as 6'-7", 6'-8", and so on, proceed as shown in the illustrative examples below.

EXAMPLE: Find the number of cubic yards of earth contained in a cellar 28' × 32', depth 6'-7".

SOLUTION: Turn to the table for the nearest depth, which is 6'-6". The result shown for 28' × 32' is 216 cu. yd. Turn to the table for depth 6'-9" (table of next greater depth). The result shown is 224 cu. yd. The difference is 8 cu. yd. As the tables are given for every 3" in depth, to find the cubic-yard contents of the odd 1" depth, take one-third of 8, which equals 2.6 cu. yd.; this plus the result shown in the 6'-6" table (216) equals 218.6 cu. yd., which is the answer. If the depth should be 6'-8", take two-thirds of 8 cu. yd., which equals 5.3 cu. yd.; this plus the result shown in the 6'-6" table (216) equals 221.3 cu. yd., which is the answer.

For odd depths of any size proceed as in the foregoing example.

TO AVERAGE DEPTHS OF CELLARS

To average the depth of cellars, apply the following rule, which is very simple:

Rule: Measure the depth of the cellar in feet and inches at the four corners, add all the results together, and divide by 4, which will give the average depth in feet and inches.

Note: Measurements are to be made from the original grade line at each corner, and care should be taken to see that the bottom of the cellar is level. Should the original surface of the ground be rolling between corners, some allowance should be made. Should an ell join onto the main excavation, figure each separately.

EXAMPLE: Find the average depth of a cellar of 30' × 34' depth, one corner 6'-4", one corner 4'-10", one corner 2'-8", and one corner 3'-6".

SOLUTION: The sum total of all corners is 17'-4", which, divided by 4, equals average depth, which is 4'-4".

EXAMPLE: Find the average depth of a cellar having the main cellar 20' × 30' and an addition 18' × 20'.

Main cellar:

One corner 7'-6"
One corner 6'-4"
One corner 4'-7"
One corner 3'-9"

Total 22'-2"

The sum total of all four corners, 22'-2", divided by 4 gives the average depth as 5'-6½".

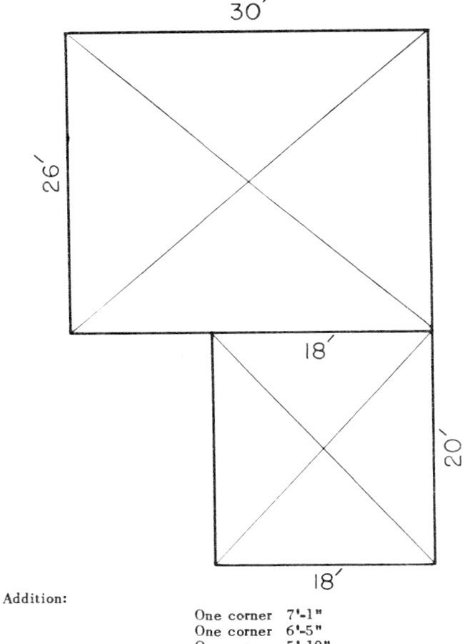

Addition:

One corner 7'-1"
One corner 6'-5"
One corner 5'-10"
One corner 5'-2"

Total 24'-6"

The sum total of all four corners, 24'-6", divided by 4 gives the average depth as 6'-1½".

PERCENTAGE OF EARTH SWELL AND SHRINKAGE

Material	Swell %	Shrinkage %
Sand or Gravel	14–16	12–14
Loam	20	17
Common Earth	25	20
Dense Clay	33	25
Solid Rock	50–75	—

HAND EXCAVATION

Hand excavation is excavation which cannot be done with mechanical equipment or in which the use of mechanical equipment would be prohibitive as far as cost is concerned. Footings for foundation walls are generally hand-excavated, as well as the area required for the footing of the chimney. Where terra-cotta drains to septic tanks are called for on the drawings, the contractor shall excavate such trenches sufficient in width to provide proper working conditions. Drain trenches are generally 3' below the lowest finished grade level and slope toward the septic tank at a pitch of ¼" to every foot of length. The excavation of the septic tanks is also generally a hand-excavation job. The excavation of the septic tank is carried 4' below the house sewer inlet. Leaching pools are usually excavated to 5' below the inlet and are dug to a diameter of 5'-8".

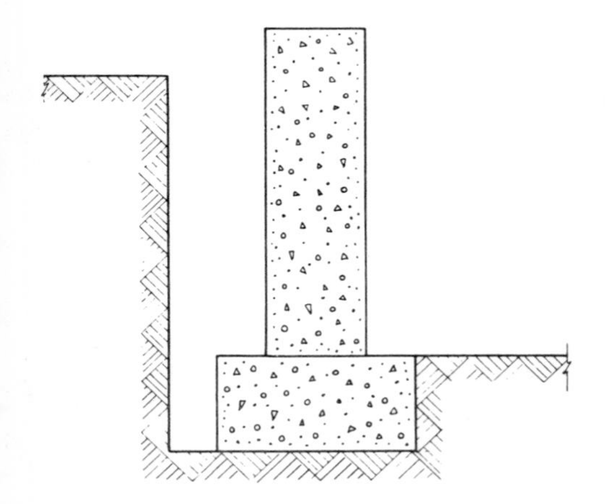

TRENCHES FOR FOOTINGS

TRENCH EXCAVATION FOR DRAIN TILE

In finding the cubic yards of earth excavation for drain tile trenches, multiply the length of the trench by its width and by its average depth. Divide by 27 to get the cubic yards. A drain tile trench leading from a building to a dry well is 75 ft. long as shown in the illustration.

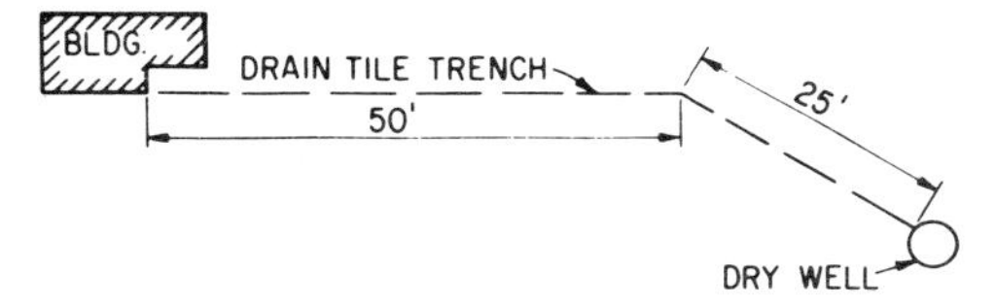

The drain tile is pitched $\frac{1}{4}$ in./ft. length. Find the cubic yards of earth excavation.

If the drain tile is pitched, the trench for the tile is similarly pitched. If the trench deepens $\frac{1}{4}$ in. for every foot of length, it is obvious that in 75 ft. there are 75/4 or $18\frac{3}{4}$ in., which is 1'-$6\frac{3}{4}$" of additional depth at the end of the trench. If the trench at the building is 3'-0' deep, the other end will be 3' + 1'-$6\frac{3}{4}$" = 4'-$6\frac{3}{4}$" deep. The average depth is therefore 3' + 4'-$6\frac{3}{4}$" = 7'-$6\frac{3}{4}$". Divide this by 2 to get the average depth of the trench.

Backfilling, which has been mentioned before, consists of filling earth around basement and foundation walls after they have set.

Grading means leveling the surface of the lot in preparation for landscaping.

VOLUME OF TRENCH EXCAVATION IN CUBIC YARDS PER 100 LIN. FT.

The following table giving various depths and trench widths will yield the cubic yards of earth to be removed for every 100 lin. ft. Find the width of the trench across the top of the table and then the depth of the trench along the right side of the table. For a trench width of 18" and a depth of 24", the excavated material is 11.1 cu. yd. per 100' of length.

CUBIC YARDS OF TRENCH EXCAVATION PER 50 FEET OF LENGTH

Depth of Trench (feet)	Width of Trench in Feet						
	2'-0"	2'-6"	3'-0"	3'-6"	4'-0"	4'-6"	5'-0"
1'-0"	3.7	4.8	5.5	6.5	7.4	8.3	9.3
1'-6"	5.5	6.9	8.3	9.7	11.1	12.5	13.9
2'-0"	7.4	9.3	11.1	13	14.8	16.7	18.5
2'-6"	9.3	11.6	13.9	16.2	18.5	20.8	23.1
3'-0"	11.1	13.9	16.5	19.4	22.2	25	27.8
3'-6"	13	16.2	19.4	22.7	25.9	29.1	32.4
4'-0"	14.8	18.5	22.2	25.9	29.6	33.3	37
4'-6"	16.7	20.8	25	29.2	33.3	37.5	41.7
5'-0"	18.5	23.1	27.8	32.4	37	41.7	46.3
5'-6"	20.4	25.5	30.6	35.6	40.7	45.8	50.9
6'-0"	22.2	27.8	33.3	38.9	44.4	50	55.6
6'-6"	24.1	30.1	36.1	42.1	48.1	54.2	60.2

MINIMUM TRENCH WIDTHS

If Depth Is	Min. Width Is	If Pipe Is	Min. Width Is
1'-0"	1'-4"	4" to 8"	1'-9"
2'-0"	1'-5"	12"	2'-6"
3'-0"	1'-6"	15"	2'-9"
4'-0"	1'-8"	18"	3'-0"
5'-0"	1'-10"	24"	3'-9"
6'-0"	2'-0"	30"	4'-3"

If trench is sheathed, add thickness of sheathing to above figures. Sheathe all trenches over 5'-0" deep for safety.

BRACING AND SHEET-PILING TRENCHES

Sheet piling is required where the soil is not self-supporting or where excavation adjoins a property line. This is estimated by the square foot, by taking the number of square feet of bank or trench walls to be braced or sheet-piled and estimating the cost of the work at a certain price per square foot for labor and lumber required.

Example:

Let us determine the quantity of lumber needed to sheet pile a trench on both sides, 28'-0" long and 7'-6" deep. The total area to be sheet piled therefore is 56' × 8' = 448 sq. ft. The lumber required can be found as follows:

2" × 8" of 8'-0" sheathing 84 pieces =	898 BF
4" × 6" of 12'-0" long stringers 9 pieces =	216 BF
4" × 6" of 3'-0" long braces 5'-0" apart 12 pieces =	72 BF
	1186 BF

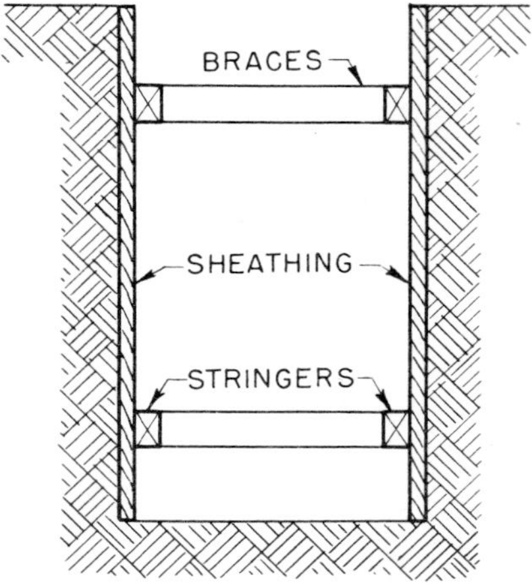

SHEET PILING
TRENCH EXCAVATION

Note: The 75 pcs. are determined by dividing the width of the planks (8") into the total length of sheet piling in inches. The length of the sheet piling is 50'-0", or 600 lin. in. Therefore, $600 \div 8 = 75$ pcs. Transposing the 75 pcs. of 2" × 8" × 8' planks into feet board measure take

$$\frac{2 \times 8 \times 8}{12} = 10.6 \times 75 = 800 \text{ f.b.m.}$$

The 6 pcs. of 4" × 6" to 16'-0" stringers or wales are determined by dividing the linear length of stringers by 16. Therefore, 4 stringers at 25 ft. long equals 100 lin. ft. $100 \div 16 = 6$ pcs. The 13 pcs. of 4" × 6" are determined by dividing the length of the trench (25') by the spacing of the braces (4'-0" on centers) and multiplying by 2, (top and bottom row). Therefore,

$$25 \div 4 = 6.25 \times 2 = 12.5 \text{ or } 13 \text{ pcs.}$$

BRACING

In excavating trenches and piers 5'-0" to 8'-0" deep, it is not always necessary to sheathe the walls solid; two or three lines of braces placed along the sides of the trench, as illustrated, will often be sufficient.

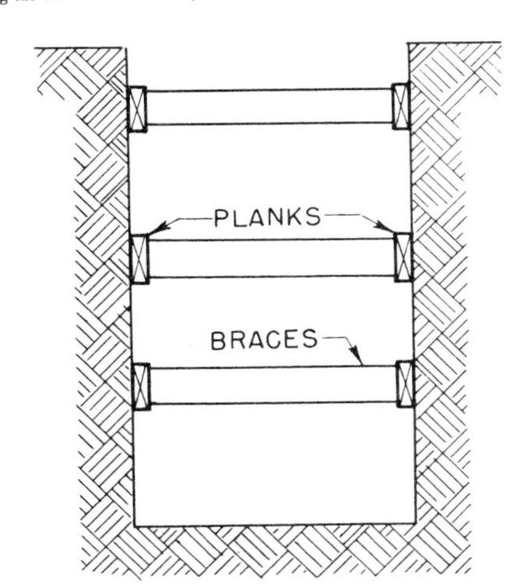

BRACING
TRENCH EXCAVATION

SHEET PILING FOR BASEMENTS AND DEEP FOUNDATIONS

On basement excavations or large piers 8'-0" to 12'-0" deep, where it is necessary to brace any one or all of the outside walls, the following is an example of the lumber required for a wall 50'-0" long, 8'-0" deep, and containing 400 sq. ft. of wall to be sheathed:

80 pcs. 2" × 8" × 10'-0" sheathing	1,067 f.b.m.
6 pcs. 6" × 8" × 16'-0" (2 lines of stringers or wales)	384 f.b.m.
10 pcs. 8" × 8" × 12'-0" top braces spaced about 5'-0" apart	640 f.b.m.
10 pcs. 8" × 8" × 6'-0" bottom braces spaced about 5'-0" apart	320 f.b.m.
10 pcs. 8" × 8" × 6'-0" stakes spaced about 5'-0" apart	320 f.b.m.
Total lumber required for 400 sq. ft.	2,731 f.b.m.
F.b.m. per sq. ft. of wall	6.8

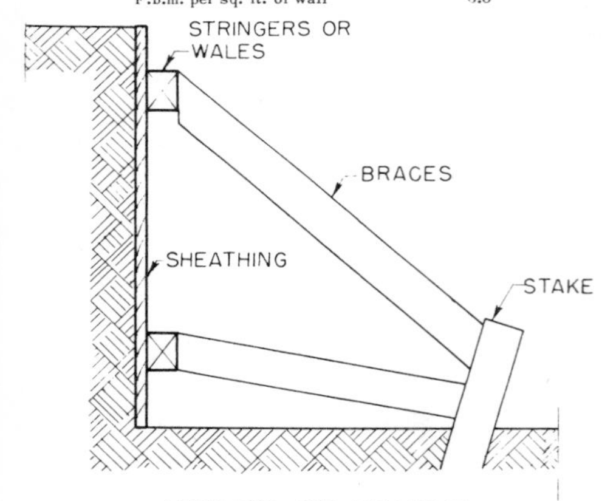

METHOD OF BRACING
DEEP BASEMENTS OR TRENCHES

Cost of lumber should be computed according to the number of times it can be used on the job.

DATA ON SHEET PILING

Size of Timber		B.F.M. Single Sheet	Number of Pieces for			
			25 Ft.	50 Ft.	75 Ft.	100 Ft.
2" × 8" Sq. edged	7'-0" Long	9.3	38	75	113	150
3" × 8" Sq. edged	8'-0" Long	16	38	75	113	150
2" × 10" Sq. edged	9'-0" Long	15	30	60	90	120
3" × 10" Sq. edged	12'-0" Long	30	30	60	90	120
3" × 12" Sq. edged	14'-0" Long	42	25	50	75	100

SIZES AND WEIGHTS OF BETHLEHEM STEEL SHEET PILING—DIMENSIONS

Section Number	Nominal Width b in.	Webb Thickness tw in.	Weight per Foot lb.	Weight per Square Foot lb.
PS 28	15	3/8	35.0	28
PS 32	15	1/2	40.0	32

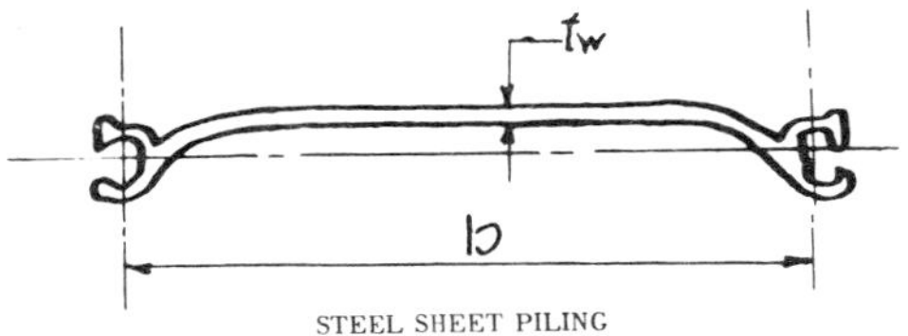

STEEL SHEET PILING

Steel sheet piling is used for supporting the soil in large excavations, deep piers, or trenches where wood piling is impractical. It is driven with a pile driver the same as wood or concrete piles.

The following table will give some of the more common sizes and weights readily available.

SIZES AND WEIGHTS OF CARNEGIE STEEL SHEET PILING

Section No.	Width, In.	Web thickness, In.	Wt., Lb. per Lin. Ft.	Wt., Lb. per Sq. Ft. of Wall
M-106	14"	3/8"	36.9	31.6
M-107	15"	3/8"	38.4	30.7
M-108	15"	1/2"	42.8	34.2
M-110	16"	31/64"	42.6	32.0
M-111	16"	3/8"	29.3	22.0
M-112	16"	3/8"	30.6	23.0
M-113	16"	1/2"	36.2	27.2
MZ-27	18"	3/8"	40.5	27.0
MZ-22	22"	3/8"	40.3	22.0

DECIMALS OF AN INCH TO MILLIMETERS

Fraction	Decimal Equivalent	Millimeters
1/32	.03125	.79375
1/16	.0625	1.58750
3/32	.09375	2.38125
1/8	.125	3.17501
5/32	.15625	3.96876
3/16	.1875	4.76251
7/32	.21875	5.55626
1/4	.25	6.35001

Multiply decimal equivalents by 25.4 to get millimeters.
(continued on page 174)

CONCRETE FOUNDATIONS, FLOORS, AND WALLS

CONCRETE FOUNDATIONS

This heading includes all column footings, basement walls, boiler foundation, area walls, and all other items of concrete below the first floor. Since basement floors undoubtedly will be installed much later than the work mentioned above, they should be listed with work done at a later period. Consequently, floors, walks, and other pavements which are being laid directly on the ground under the heading "Cement Finish" will be listed later. There are three parts to each item of concrete work, and these parts, which must be listed separately, are concrete, formwork, and reinforcing steel.

Concrete footings and concrete foundation walls are estimated by the cubic yard. Length of footings in feet, multiplied by width in feet, multiplied by depth in feet, and divided by 27 will give the amount of cubic yards in the footing.

HOW TO ESTIMATE MATERIALS REQUIRED FOR CONCRETE FLOORS, SIDEWALKS, DRIVEWAYS AND OTHER FLAT SURFACES OF A GIVEN THICKNESS

To estimate the concrete materials required for a 6-in. thick 30 x 60 ft. concrete floor: In the following table, 100 sq. ft. of floor 6-in. thick (1:4 mix) requires 12.5 sacks of Portland cement and 57 cu. ft. of sand-gravel aggregate. The area of the floor is 30 x 60 ft. = 1800 sq. ft. of surface; 1800 ÷ 100 = 18. Therefore, total amounts of materials required are found as follows:

18 x 12.5 = 225 sacks of cement

18 x 57 = 1036 cu. ft. or 38 cu. yds of sand-gravel aggregates

HOW TO ESTIMATE MATERIALS REQUIRED FOR 100 SQ. FT. OF CONCRETE OF THICKNESSES SHOWN

Thickness of Concrete inches	Amount of Conc. Cu. Yds.	Proportions			
		1:3½ Mix		1:4 Mix	
		Cement Sacks	Sand-Gravel cu. ft.	Cement Sacks	Sand-Gravel cu. ft.
3	0.93	7.2	28	6.2	28
4	1.22	9.6	38	8.4	38
5	1.52	12.1	48	10.5	48
6	1.85	14.3	57	12.5	57
7	2.15	16.7	67	14.6	67
8	2.48	19.1	76	16.6	76
10	3.07	24.0	95	20.8	95
12	3.70	28.6	114	25.0	114

Sand-gravel quantities required have been increased by about 15 per cent to allow for volume variation and waste. 1 cu. yd. of sand-gravel weighs about 3,200 lbs.

FORMWORK

Formwork is figured per square foot of contact of the forms against the concrete. Forms must be strong, watertight, and straight to produce a well-built straight wall. They must be braced very securely so that the pressure of the poured concrete will not make the forms bulge. The pressure which has to be resisted by all formwork amounts to about 150 lb. per sq. ft. for each foot of height. Forms for footings should be used in all instances except where a hard clay, which can be cut with the shovel, is encountered. The quantity of forms for wall footings is determined by measuring the outside perimeter of the wall and footing, thus establishing the amount of contact area between wall and footing. After this has been established, the amount of contact area is found for the inside. The two items are then added together.

Suppose we had established the requirements of 120 sq. ft. of forms for a footing and the material used were 1" × 6" boards, it would be necessary to multiply the area by 2 for both sides of the footing. Since this lumber is bought in board feet measure, it is necessary to change the square feet into feet board measure, thus:

$$\frac{120 \times 2 \times 6}{12} = 120 \text{ f.b.m.}$$

MATERIALS FOR FORMWORK

It is best to use tongue-and-groove boards for forms provided the forms are made up in certain sized panels. Use square-edge lumber if each piece of lumber is set in place. It is much easier to dismantle square-edge lumber than the tongue-and-groove panels. Plywood is excellent for panels because it leaves a surface that requires very little or no rubbing. Steel forms are good for foundation walls and round columns.

For form sheathing it is best and perhaps most economical to use 1" × 8" boards, which require 2 nails at each nailing point. Boards of 1" × 6" wide space the nails too close, and 1" × 10" boards require 3 nails, and the middle nail may sometimes split the board.

For vertical formwork, 2" × 6" are favored. Use two 2" × 6" for walers.

For 1,000 sq. ft. of form contact area, allow 30 lb. of No. 9 annealed wire nails.

Form boards may be used again for roof sheathing, wall sheathing, or underflooring. Form studs may also be re-used for wall studs.

QUANTITY OF CONCRETE FOR VARIOUS WALL THICKNESSES

Wall Thickness Inches	Cu. Ft. of Conc. Per Sq. Foot of Wall	Cu. Yds. of Conc. Per Sq. Foot of Wall	Wall Thickness Inches	Cu. Ft. of Conc. Per Sq. Foot of Wall	Cu. Yds. of Conc. Per Sq. Foot of Wall
3	.25	.0092	16	1.33	.049
4	.33	.0122	18	1.5	.055
6	.50	.0185	24	2.0	.074
8	.67	.025	30	2.5	.092
10	.83	.031	36	3.0	.111
12	1.00	.037	42	3.5	.13
14	1.17	.043	48	4.0	.148

QUANTITY OF CONCRETE FOR VARIOUS SIZES OF FOOTINGS

Footing Dimensions Inches		Cu. Ft. of Conc. Per Foot of Length	Cu. Ft. of Conc. Per 10 Ft. of Length	Cu. Yds. of Conc. Per Foot of Length	Cu. Yds. of Conc. Per 10 Ft. of Length
Height	Width				
6	6	.25	2.5	.0092	.092
6	8	.33	3.33	.0122	.122
6	10	.416	4.16	.0154	.154
8	8	.436	4.36	.0161	.161
8	10	.549	5.49	.0203	.203
8	12	.667	6.67	.0247	.247
10	10	.689	6.89	.0255	.255
10	12	.834	8.34	.0308	.308
12	12	1.00	10.00	.0370	.370
12	14	1.167	11.67	.0432	.432
12	16	1.33	13.33	.0492	.492
12	18	1.5	15.00	.0555	.555
12	20	1.67	16.76	.0618	.618
12	22	1.83	18.34	.0677	.677
12	24	2.00	20.00	.0740	.740
14	24	2.33	23.33	.0863	.863
14	26	2.64	26.45	.0977	.977
16	26	2.88	28.82	.1066	1.066
16	28	3.09	30.99	.1144	1.144
18	30	3.75	37.50	.1388	1.388

FORMS FOR BEAMS AND GIRDERS

To determine the number of square feet of formwork for reinforced beams and girders, add the three sides of the beam and girder, and multiply by the length. The dimensions of the beam or girder are taken from the underside of the slab to the bottom of the beam and girder.

For example: In the following illustration add 9" + 9" + 9" = 27", or 2' − 3", or 2.25'.

2.25' × the length of beam = sq. ft. of forms.

It is considered good practice to estimate formwork for beams and girders separately. Forms are generally of 1" material. Where a smooth surface is desired, plywood should be used in place of sheathing.

The following tables will give the square feet of forms for beam depths ranging from 7" to 66" and widths from 7" to 16".

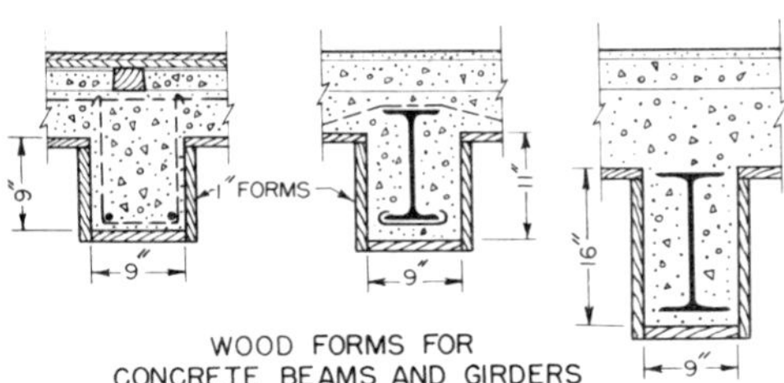

WOOD FORMS FOR CONCRETE BEAMS AND GIRDERS

ESTIMATING FORMS FOR CONCRETE SLAB

The form boards are of 1" lumber supported by 2" × 8" joists spaced 24" O.C. (on center). The ends of the joists are supported by 2" × 6" vertical members also spaced 24" O.C. The ribbons of 1" × 6" material further support the joists. Assuming a bay length of 20'-0", the following quantities of lumber are required:

Area of forms (supporting slab) 20 × 8 = 160 sq. ft. plus 15% waste	= 184 B.F.
Joists, 20' ÷ 2 = 10 joists plus 1 = 11 joists − 11 pieces at 2" × 8" × 8"	= 116 B.F.
Ribbons, 2 × 20' = 40 lin. ft. of 1" × 6"	= 20 B.F.
Vertical 2" × 6" members, 22 pieces 2' − 3" long, or 2.25 × 22	= 50 B.F.
	Total = 370 B.F.

Since 1 bay has 160 sq. ft., then

370 ÷ 160 = 2.31 B.F. per sq. ft. of surface.

Note: Forms for beams should be figured separately.

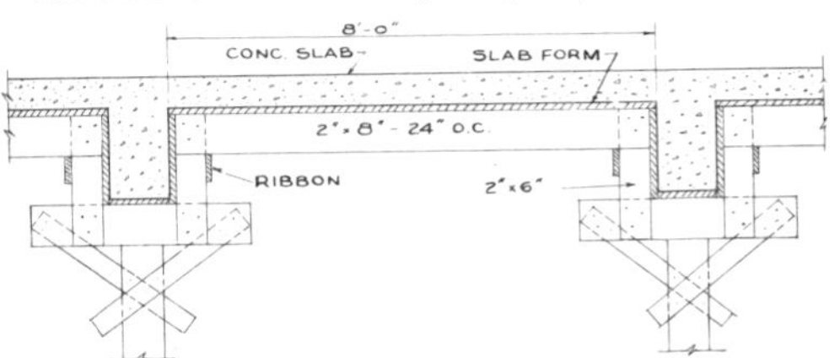

REINFORCING STEEL

Concrete footings, in many instances, require reinforcing rods to give additional strength to the footing. These steel rods are figured by their actual length and their diameters. All rods of equal diameter are added up and the total multiplied by the weight per foot to arrive at the total tonnage. In recent years considerable changes in reinforcing-bar design have evolved. The total weight of the bars remains the same as previously, but the actual diameter of the bar is slightly less. For this reason, instead of calling the bars ¼", ⅜", etc., they will hereafter be known as Nos. 3, 4, 5, etc.

For example: Find the total tonnage of ¾"-diameter reinforcing rods required in the footing of the following illustration. Since reinforcing rods are allowed to overlap at the corners, the 36' dimension, plus 1' for footing extension, 6" on either end, or a total of 37', is multiplied by 6, for two lengths of wall having three reinforcing rods each. Add 1' to the 22' dimension, or 23', and multiply by 6.

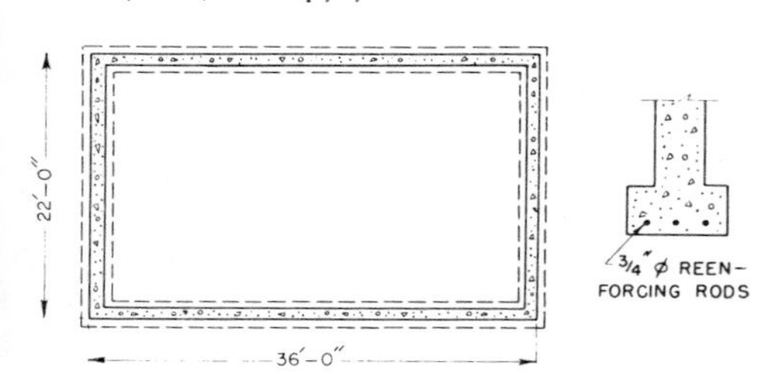

ESTIMATING REENFORCING STEEL

If, therefore, the perimeter of the footing is found, the total length may be multiplied by 3 to get the total linear feet of rods. In the table on weights of steel bars, the weight per foot can be found and multiplied by the total feet. The result is so many pounds. The pounds are divided by 2,000 (2,000 lb. = 1 ton) to get the tonnage. Thus 37 + 37 + 22 + 22 × 3 = 354 lin. ft. per ft. of ¾" reinforcing rod weighs 1.5 lb.

$$354 \times 1.5 = 531 \text{ lb.}$$
$$531 \div 2{,}000 = .265 \text{ ton}$$

STANDARD SIZES AND WEIGHTS OF CONCRETE REINFORCING BARS

Former Bar Designation, In.	New Bar Designation, No.	Unit Weight, Lb. per Ft.	Diameter, In.	Cross-sectional Area, Sq. In.	Perimeter, In.
¼ round	2b	.167	.250	.05	.786
⅜ round	3	.376	.375	.11	1.178
½ round	4	.668	.500	.20	1.571
⅝ round	5	1.043	.625	.31	1.963
¾ round	6	1.502	.750	.44	2.356
⅞ round	7	2.044	.875	.66	2.749
1 round	8	2.670	1.000	.79	3.142
1 square	9	3.400	1.128	1.00	3.544
1⅛ square	10	4.303	1.270	1.27	3.990
1¼ square	11	5.313	1.410	1.56	4.430

AMOUNT OF CONCRETE IN ONE-BAG BATCH FOR DIFFERENT PROPORTIONS OF SAND AND GRAVEL

Proportions	Amount of Concrete in One-bag Batch	
	Cu. Ft.	Cu. Yd.
1:1½:3	3.53	.131
1:2:3	3.90	.145
1:2:3½	4.22	.156
1:2:4	4.50	.167
1:2½:4	4.88	.181
1:2½:4½	5.17	.192
1:2½:5	5.40	.202
1:3:5	5.81	.215
1:3:5½	6.11	.226
1:3:6	6.38	.236

SUITABLE MIXTURES FOR VARIOUS CONCRETE CONSTRUCTION PROJECTS

	One-mixture Concrete Construction	Two-course Slab
Foundation walls and footings	1:2¾:4 1:3:5 1:2½:5	
Basement walls	1:2½:4 1:2¾:4	
Basement walls—waterproof	1:2¼:3 1:2½:3½ 1:2½:4½	
Retaining walls	1:2:3½ 1:3:5	
Steps	1:2¼:3	
Lintels	1:2:4	
Swimming pools	1:2:3 1:2½:3	
Barnyard pavements	1:3:5	
Beam filling	1:3:4	
Fence posts	1:1:1½ 1:1¾:2 1:2:3 1:2¼:2½	

SUITABLE MIXTURES FOR VARIOUS CONCRETE CONSTRUCTION PROJECTS (*continued*)

	One-mixture Concrete Construction	Two-course Slab	
		Base	Top
Floors:			
One-course	1:1¾:4 1:2½:3		
Heavy-duty, one-course	1:1:2 1:1¼:2		
Farm buildings:			
One-course	1:2¼:3		
Two-course		1:2¼:3 1:2½:4	1:1½ 1:2
Driveways:			
One-course	1:2:3½ 1:2¼:3 1:2½:3		
Two-course		1:2¼:3 1:2½:4	1:1½ 1:2 1:1:1½
Sidewalks:			
One-course	1:2¼:3 1:2½:4		
Two-course		1:2¼:3 1:2½:4	1:1½ 1:2

MATERIAL REQUIRED FOR 100 SQ. FT. OF CONCRETE FLOOR BASE

Thickness, In.	Proportions								
	1:2:3			1:2:4			1:2½:5		
	Cement, Bbl.	Sand, Cu. Yd.	Pebbles, Cu. Yd.	Cement, Bbl.	Sand, Cu. Yd.	Pebbles, Cu. Yd.	Cement, Bbl.	Sand, Cu. Yd.	Pebbles, Cu. Yd.
3	1.62	.48	.71	1.38	.41	.82	1.15	.43	.85
3½	1.89	.56	.83	1.61	.48	.96	1.35	.50	1.00
4	2.16	.64	.95	1.84	.55	1.10	1.54	.56	1.23
4½	2.43	.72	1.07	2.07	.62	1.24	1.73	.63	1.26
5	2.68	.80	1.19	2.31	.69	1.37	1.92	.70	1.41

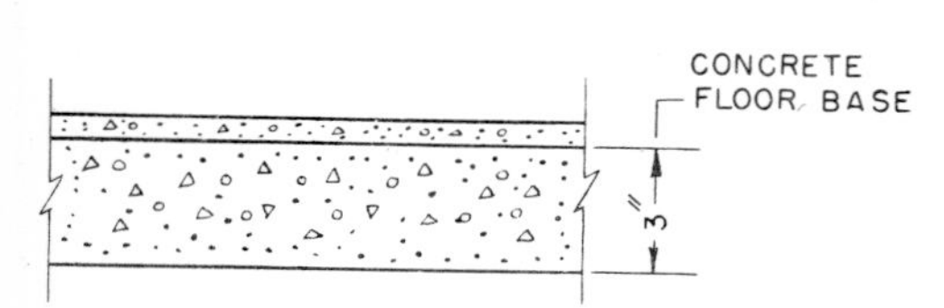

To determine the quantity of cement (barrels), sand (cubic yards), and pebbles (cubic yards) in a concrete floor base, measuring 10'-0" × 15'-0", proceed as follows:

$$10 \times 15 = 150 \text{ sq. ft.}$$

For 100 sq. ft. of 3" floor base, in the above table, the quantity of

Cement = 1.62 bbl.
Sand = .48 cu. yd.
Pebbles = .71 cu. yd.

Therefore, multiply

1.5 × 1.62 = 2.43 bbl. of cement
1.5 × .48 = .72 cu. yd. of sand
1.5 × .71 = 1.06 cu. yd. of pebbles

For 225 sq. ft. of floor base, multiply the values in the table by 2.25.

AMOUNTS OF MATERIALS FOR 100 SQ. FT. OF WALL AREA

Thickness of Wall, In.	Mixture					
	1:2½:5			1:2:4		
	Cement, Bbl.	Sand, Cu. Yd.	Pebbles, Cu. Yd.	Cement, Bbl.	Sand, Cu. Yd.	Pebbles, Cu. Yd.
6	2.30	.85	1.70	2.70	.83	1.66
8	3.08	1.13	2.26	3.70	1.10	2.20
10	3.85	1.41	2.82	4.63	1.37	2.70
12	4.60	1.70	3.40	5.56	1.66	3.30
15	5.76	2.12	4.24	6.93	2.06	4.12
18	6.90	2.55	5.10	8.34	2.49	4.98

QUANTITIES FOR 100 SQ. FT. WEARING SURFACE OR TOPCOAT

Thickness, In.	Proportions				
	1:2		1:1:1		
	Cement, Bbl.	Sand, Cu. Yd.	Cement, Bbl.	Sand, Cu. Yd.	Pebbles, Cu. Yd.
½	.51	.15			
¾	.75	.23			
1	1.00	.29	1.00	.15	.15
1¼	1.26	.37	1.26	.19	.19
1½	1.51	.45	1.51	.23	.23
2	2.00	.59	2.00	.30	.30

SURFACE-COVERING CAPACITY OF MORTAR FROM ONE BAG OF CEMENT

Mixture of Parts by Volume		Thickness of Coat				
		¼"	⅜"	½"	¾"	1"
Cement	Sand	Sq. Ft.	Sq. Ft.	Sq. Ft.	Sq. Ft.	Sq. Ft.
1	1	66	44	33	22	16
1	1½	84	56	42	28	21
1	2	101	67	50	33	25
1	2½	118	78	59	39	29
1	3	136	90	68	45	34
1	3½	153	102	76	51	42
1	4	171	113	85	57	38

RECOMMENDED THICKNESSES OF CONCRETE SLABS IN INCHES

Basement floors for dwellings	4
Private garage floors	4 to 5
Porch floors	4 to 5
Stock barn floors	5 to 6
Poultry-house floors	4
Hog-house floors	4
Milkhouse floors	4
Granary floors	5
Impliment-shed floors	6
Tile floor bases	2½
Driveways and approaches	6 to 8
Sidewalks	4 to 6

COLORS TO BE USED IN CONCRETE FLOOR FINISH

Color Desired	Commercial Names of Colors for Use in Cement	Lb. of Color Required per Sack of Cement to Obtain	
		Light Shade	Medium Shade
Grays, blue-black, and black	Germantown Lampblack*,	½	1
	or, carbon black*, or,	½	1
	black oxide of Manganese*,	1	2
	or mineral black*	1	2
Blue shade	Ultramarine blue	5	9
Brownish red to dull brick red	Red oxide of iron	5	9
Bright red to vermilion	Mineral turkey red	5	9
Red sandstone to purplish red	Indian red	5	9
Brown to reddish brown	Metallic brown (oxide)	5	9
Buff, Colonial tint, and yellow	Yellow ocher, or	5	9
	yellow oxide	2	4
Green shade	Chromium oxide, or	5	9
	greenish-blue ultramarine	6	

*Only first-quality lampblack should be used. Carbon black is light and requires very thorough mixing. Black oxide or mineral black is probably most advantageous for general use. For black, use 11 lb. oxide per sack of cement.

SQUARE FEET COVERAGE OF VARIOUS QUANTITIES OF CONCRETE AT VARIOUS THICKNESSES

Cu. Yds. of Conc.	1" Thick	1½" Thick	1¾" Thick	2" Thick	2½" Thick	3" Thick	3½" Thick	4" Thick
1	324	216	185	162	130	108	93	81
1½	486	324	278	243	194	162	138	121.5
2	648	432	361	324	259	216	185	162
2½	810	540	454	405	324	270	231	202.5
3	972	648	547	486	389	324	277	243
3½	1134	756	640	567	454	378	323	283.5
4	1296	864	733	648	519	432	369	324
4½	1458	972	826	729	584	486	415	364.5
5	1620	1080	919	810	649	540	461	405
5½	1780	1188	1012	891	714	594	507	445.5
6	1942	1296	1105	972	779	648	553	486
6½	2164	1404	1198	1053	844	702	599	526.5

FEET BOARD MEASURE OF LUMBER REQUIRED FOR FORMS FOR 1 SQ. FT. OF CONCRETE SIDEWALK

Size Lumber	Width of Sidewalk, Ft.								
	2'	3'	4'	5'	6'	7'	8'	9'	10'
2" × 4"	.90	.625	.50	.375	.30	.25	.25	.20	.20
2" × 6"	1.25	.875	.625	.50	.40	.375	.33	.30	.25
2" × 8"	1.66	1.125	.875	.66	.60	.50	.40	.375	.33

Labor placing wood forms and screeds for concrete sidewalks and floors. Under average conditions, a man should place and level 200 to 240 lin. ft. of 2" × 4", 2" × 6", or 2" × 8" screeds per 8-hr. day, or it requires 3, 6 hr. per 100 lin. ft. of screed or form.

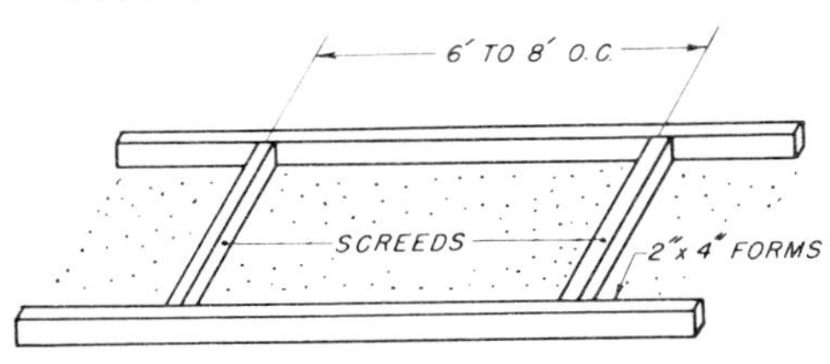

FORMS OR SCREEDS FOR CONC. WALKS

BRICK MASONRY

HOW TO ESTIMATE COMMON BRICK

In present-day estimating of brickwork, the cubic feet of wall are determined. This means multiplying the length by the height by the thickness of wall. All openings, such as windows, doors, or other openings, should be deducted in order to get the net cubic feet of wall. This is then multiplied by the number of brick per square foot of wall. Many contractors in the past multiplied the net cubic feet of wall by 20 brick per cubic foot. This number already included about 5% for waste.

This method, although close, is not entirely satisfactory in present-day estimating, where competition is a factor to be considered, and especially when contractors are figuring on a 5 to 10% margin of profit.

In order to get a more accurate figure, the mortar joint will have to be considered. Mortar joints vary from ⅛" to ⅝", with ½" the average width.

For example: A brick measures 8" long by 2¼" high. By adding ¼" for the vertical, or end, mortar joint and ½" for the horizontal, or bed, mortar joint, the brick's size can then be taken as 8¼" long by 2¾" high. Therefore 8¼" × 2¾" = 22¹¹⁄₁₆ sq. in. on the face of the brick.

To obtain the number of brick per square foot of wall, divide 144 by 22¹¹⁄₁₆, and the result is 6.35, or 6⅓ brick per square foot of 4" wall. If the wall is 8" thick, or 2 brick thick, multiply 2 × 6⅓ = 12⅔ brick per square foot. If a 12" wall is called for, or 3 brick thick, multiply 3 × 6⅓ = 19 brick per square foot of wall. This can be done for any thickness of wall if the number of brick thickness and the vertical and horizontal mortar joints are known.

The following table will yield the number of common brick (8" × 2¼" × 3¾") required for 1 sq. ft. of brick wall of any thickness, with the mortar joints specified. Brick walls are generally designated on the plans as 4" or 4½", 8" or 9", 12" or 13", increasing by 4" or 4½" in width. This variation in thickness need not be considered because a wall marked 13" on the plans does not require any more brick than a 12" wall, as both walls are 3 brick thick.

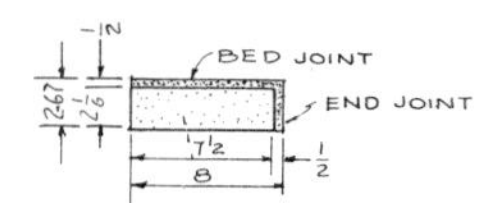

NUMBER OF MODULAR BRICK REQUIRED FOR
1 SQ. FT. OF WALL.
Actual Brick Size 7½ × 2⅙ × 3½ (Nominal Size 8" × 4" × 2⅔")

Wall Thickness	Width of Vertical or End Joints—Same as Horizontal or Bed Joint						
	⅛	¼	⅜	½	⅝	¾	⅞
4	7.72	7.71	7.46	6.75	6.42	6.00	5.71
8	15.44	15.42	14.92	13.50	12.84	12.00	11.42
12	23.16	23.13	22.38	20.25	19.26	18.00	17.13
16	30.88	30.84	29.84	26.00	25.68	24.00	22.84
20	38.6	38.55	37.30	33.75	32.10	30.00	28.55
24	46.32	46.26	44.76	40.50	38.52	36.00	34.26

IMPORTANT FACTORS TO BE CONSIDERED IN LAYING BRICK

To achieve the most economical results in laying brick, it is best to use screened sand and mortar so that bricklayers will not waste time picking out pebbles. In mixing mortar in tubs, it is best to use seamless metal ones because they do not leak or catch the trowel. Batch mixers are economical if the job is large enough. When brick is delivered, it should be stacked instead of dumped. This makes for faster handling, reduces breakage and chipping, and requires less storage space. Bricks should be picked up with tongs and placed in a special flat wheelbarrow and not in the barrow used for mortar. Wh.n brick is delivered, have it stacked within reasonable wheeling distance. The best bricklayers should be placed at corners.

PROPER JOINTS IN BRICKWORK

Mortar joints must be made carefully. Bed joints should be struck full and be spread thick to one inch or more. Bed joints are important from the standpoints of bonding bricks together, creating equal pressure throughout a wall. It is customary to run the point of the trowel along the mortar to make a furrow. It is not advisable to spread bed mortar more than a distance of four to five brick lengths in advance of laying. Head joints like bed joints must be completely filled with mortar.

NUMBER OF DOUBLE-SIZE COMMON BRICK REQUIRED FOR 1 SQ. FT. OF BRICK WALL
(Brick size 8" × 5" × 3¾")

Thickness of Wall	Number of Brick Thick	Width of Horizontal, or Bed, Mortar Joints					
		⅛"	¼"	⅜"	½"	⅝"	¾"
4" or $4\frac{1}{2}$"	1	3.41	3.34	3.24	3.2	3.1	3.03
8" or 9"	2	6.82	6.68	6.48	6.4	6.2	6.06
12" or 13"	3	10.23	10.02	9.72	9.6	9.3	3.09
16" or 17"	4	13.64	13.36	12.96	12.8	12.4	12.12
20" or 21"	5	17.05	16.70	16.20	16.0	15.5	15.15
24" or 25"	6	20.46	20.04	19.44	19.2	18.6	18.18

CUBIC FEET OF MORTAR REQUIRED PER 1,000 BRICK*
(No allowance for waste)

Joint Thickness	Various Wall Thicknesses					
	4"	8"	12"	16"	20"	24"
⅛"	2.9	5.6	6.5	7.1	7.3	7.5
¼"	5.7	8.7	9.7	10.2	10.5	10.7
⅜"	8.7	11.8	12.9	13.4	13.7	14.0
½"	11.7	15.0	16.2	16.8	17.1	17.3
⅝"	14.8	18.3	19.5	20.1	20.5	20.7
¾"	17.9	21.7	23.0	23.6	24.0	24.2
⅞"	21.1	25.1	26.5	27.1	27.5	27.8
1"	24.4	28.6	30.1	30.8	31.2	31.8

*Courtesy of the Quality Lime Institute, Philadelphia, Pa.

MORTAR MIXES REQUIRED TO LAY 1,000 COMMON BRICK

Kind of Mortar	Lime, Lb.	Cement, Bbl.	Sand, Cu. Ft.
1:3 lime mortar	180	......	18
2:1:9 lime-cement mortar	115	.50	18
1:3 cement mortar, 10% lime	16	1.50	18
Brixment cement mortar	......	1.25	15
Masonry cement mortar	......	1.25	15
Kosmortar	......	1.25	15

NUMBER OF ROMAN BRICK REQUIRED FOR 1 SQ. FT. OF WALL

Actual Brick Size 12" × 1½" × 4"

Wall Thickness	Width of Vertical or End Joint—Same as Horizontal or Bed Joint						
	⅛	¼	⅜	½	⅝	¾	⅞
4	7.31	6.71	6.20	5.76	5.367	5.01	4.70
8	14.62	13.42	12.40	11.52	10.734	10.02	9.40
12	21.93	20.13	18.60	17.28	16.101	15.03	14.10
16	29.24	26.84	24.80	23.04	21.468	20.04	18.80
20	36.55	33.55	31.00	28.80	26.835	25.05	23.50
24	43.86	40.26	37.20	34.56	32.202	30.06	28.20

NUMBER OF ENGLISH BRICK REQUIRED FOR 1 SQ. FT. OF WALL

Actual Brick Size 9" × 3" × 4½"

Wall* Thickness	Width of Vertical or End Joints—Same as Horizontal or Bed Joints						
	⅛	¼	⅜	½	⅝	¾	⅞
4½ or 1 Brick	5.14	4.71	4.54	4.33	4.12	3.28	3.81
9½ or 2 Brick	10.28	9.42	9.08	8.66	8.24	6.56	7.62
14½ or 3 Brick	15.42	14.13	13.62	12.99	12.36	9.84	11.43
21½ or 4 Brick	20.56	18.84	18.16	17.32	16.48	13.12	15.24
24½ or 5 Brick	25.70	23.55	22.70	21.65	20.60	16.40	19.05
29½ or 6 Brick	30.84	28.26	27.24	25.98	24.72	19.68	22.86

*Note Wall thicknesses include ½" motar joints between brick.

NUMBER OF BRICKS LAID PER HOUR

Walls	4" Brick	8" Brick	12" Brick
Common brick:			
Flush joint	80	120	150
Struck joint	72	110	140
Face brick:			
Running bond, cut joint	70	115	150
Running bond, V joint	60	110	145
Running bond, concave joint	60	110	145
Running bond, raked joint	50	90	125
Enameled brick	20 to 30		

PERCENTAGES ADDED TO FACE BRICK FOR VARIOUS BONDS

Type of Bond	Full Headers	Percentage to be Added
Common	Every 5th Course	20% or 1/5
Common	Every 6th Course	16.7% or 1/6
Common	Every 7th Course	14.3% or 1/7
English	Every 6th Course	16.7% or 1/6
English	Every other Course	50% or ½
Flemish	Every 6th Course	5.6% or 1/18
Flemish	Every Course	33.3% or 1/3

POINTING OF OLD BRICK WALLS

Brick walls that have been exposed to the weather for a long time may require repointing to make them more water and airtight. The old and loose mortar in the joints must be cleaned out to a depth of at least ¼ inch. The old mortar can be loosened with the small end of the mason's hammer, then scraped out with pointing tools made specially for this purpose. A thin chisel and hammer can also be used. Finally a stiff brush should be used to brush out all loose particles. New mortar is then forced into the joints so that the cavity is filled completely and the mortar sticks in place. It may be first necessary to wet the joints before the new mortar is applied. A repointed wall can be made most attractive.

PAINT FOR EXTERIOR MASONRY WALLS

The best paint for brick, concrete, or stone exterior walls is Portland cement paint and paints with a base of latex, vinyl or acrylic emulsions. Portland cement paint fills porous areas in the masonry and thereby waterproofs the surface. It consists of Portland cement and lime as the white paint solids, with water as the vehicle. Color pigments, exterior pigments, and water repellents are often added. Portland cement paint must be applied to a damp surface with a stiff brush. The paint with a base latex, vinyl or acrylic emulsions can be applied with brush, roller or spray gun.

IMPORTANT DATA ON VARIOUS TYPES OF BRICK

BUILDING BRICK—There are three classifications or grades of brick based on their resistance to weather conditions.

GRADE SW—This brick is used where a very high degree of frost is encountered or where the brick may be frozen when permeated with water. These brick may be used for foundation courses and retaining walls.

GRADE MW—This brick is used where freezing temperatures occur but unlikely to be permeated with water. This brick may be used on the face of a wall above grade. Under this use the brick is not likely to be permeated with water. It can dry out easily.

GRADE NW—This type of brick is intended for back-up or for interior walls, or if exposed to the outside where no frost action occurs.

Note: The word "permeated" as referred to in the ASTM (American Society for Testing Materials) means the complete saturated of a brick with water where moisture is drawn through the brick by capillary action,

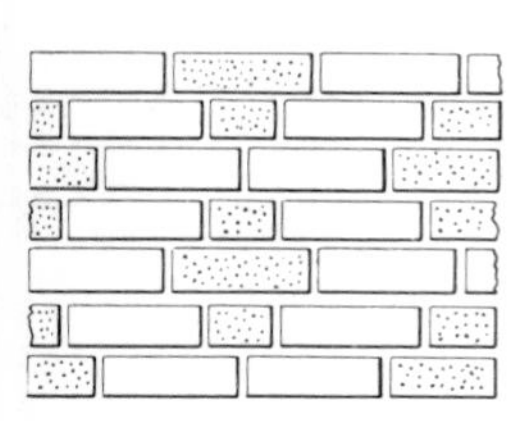

ALTERNATE FULL HEADERS EVERY 6TH COURSE
7.15 BRICK PER SQ. FT.
FLEMISH CROSS BOND

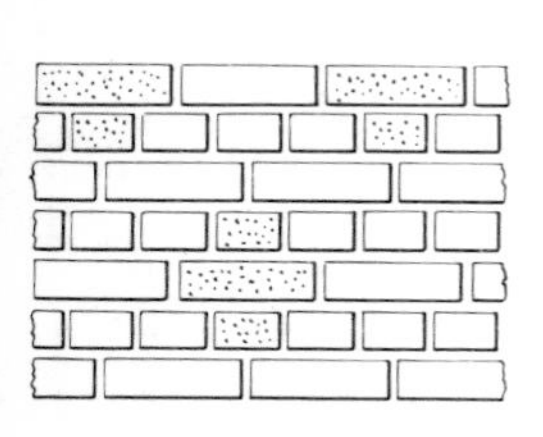

CONTINUOUS FULL HEADERS EVERY 6TH COURSE
7.88 BRICK PER SQ. FT.
ENGLISH CROSS BOND

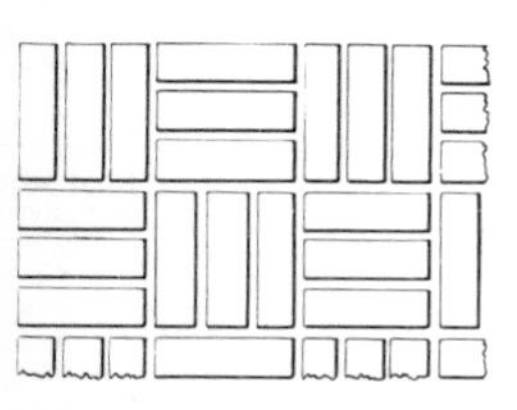

6.75 BRICK PER SQ. FT.
BASKET PATTERN

ELEVATIONS OF VARIOUS BRICK BONDS

wetting the brick from face to face. When a brick is completely immersed in water it will become saturated within a 24 hour period.

FACING BRICK—There are two grades of facing brick based on, 1. resistance to weather conditions and 2. factors based on appearance of the finished brick. Regarding the resistance to weather conditions the same is true of these brick as for building brick grades SW and MW.

Regarding factors based on appearance there are the following types:

FBX—Where a high degree of mechanical perfection is desired, either on inside or outside walls.

FBS—Where wide color ranges are desired. Used for both interior and exterior masonry walls.

FBA—A non-uniform size brick intended for architectural affect and variations.

ELEVATIONS OF VARIOUS BRICK BONDS

ESTIMATING QUANTITIES OF BRICK FOR CHIMNEYS

To find the number of brick required to build a chimney or any section of a chimney constructed wholly from one kind of brick, multiply the number of brick in each course by the number of courses to be constructed.

To find the number of common brick required to build a chimney or any section of a chimney constructed with both common and face brick, multiply the number of common brick used in each course by the number of courses to be constructed.

To find the number of face brick, multiply the number of face brick in each course by the number of courses to be constructed. After the total number of common brick or face brick is found, add 2% to allow for waste and breakage.

EXAMPLE: A brick outside chimney having a total height of 33'-0" from footing to 8' above the roof contains one 8" × 12" flue. The brick used are regular size, 2¼ × 3¾ × 8. The chimney extends out 12" from the outside wall. Both face and common brick are used in the construction of the chimney. Find the number of each kind of brick used.

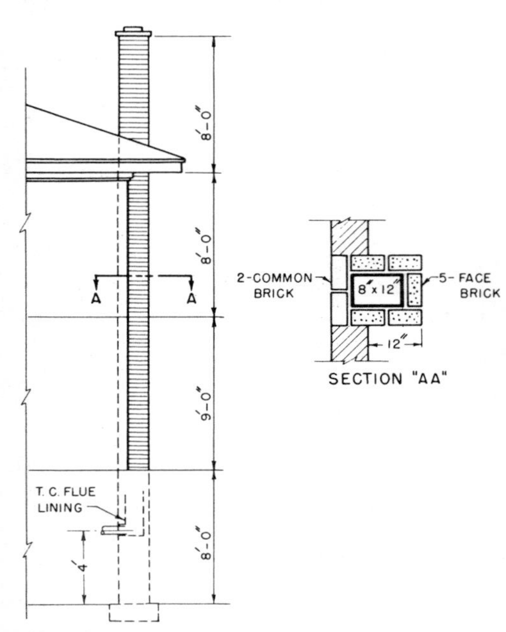

ESTIMATING QUANTITIES OF BRICK FOR CHIMNEY

SOLUTION

1. 33 − 8 = 25', height of lower chimney section.
   12 × 25 = 300", height of lower section.
   300 ÷ 2¾ = 109, number of courses of brick in lower section.

2. 12 × 8 = 96", height of upper section.
   96 ÷ 2¾ = 35, number of courses of brick in upper section.

3. 109 × 5 = 545, number of face brick required for lower section.
   109 × 2 = 218, number of common brick required for lower section.
   35 × 7 = 245, number of face brick required for upper section.

4. 545 + 245 = 790, total number of face brick required.
   2% × 790 = 15.80 or 16, number of additional face brick allowed for waste and breakage.
   790 + 16 = 806, total number of face brick required.
   2% × 218 = 4.36 or 5, number of additional common brick allowed for waste and breakage.
   218 + 5 = 223, total number of common brick required.

NUMBER OF FACE AND COMMON BRICK REQUIRED
PER CROSS-SECTIONAL AREA OF CHIMNEYS
(Chimney extends 8" from wall.)

Number of Flues	Size of Flue	Face Brick	Common Brick	Total
1	4" × 8"	3½	1½	5
1	8" × 8"	3½	2½	6
1	8" × 12"	4	3	7
1	12" × 12"	4	4	8
1	12" × 16"	4½	4½	9
1	16" × 16"	4½	5½	0
2	8" × 12"	4½	4½	9
2	8" × 12"	5	6½	11½
2	12" × 12"	5½	5½	11
2	8" × 12" 12" × 12"	5½	7	12½
2	12" × 12"	6	7½	13½
2	8" × 12"	6	6	12
3	8" × 12"	8	9	17
3	8" × 12"	6½	9½	16
3	8" × 12" 8" × 12" 12" × 12"	6½	8	14½
3	8" × 12" 12" × 12" 12" × 12"	7	8½	15½
3	12" × 12"	7½	9	16½
3	12" × 12"	8	11	19

NUMBER OF FACE AND COMMON BRICK REQUIRED PER CROSS-SECTIONAL AREA OF CHIMNEYS
(Chimney extends 12" from wall.)

Number of Flues	Size of Flues	Face Brick	Common Brick	Total
1	4" × 8"	4½	½	5
1	8" × 8"	4½	1½	6
1	8" × 12"	5	2	7
1	12" × 12"	5	3	8
1	12" × 16"	5½	3½	9
1	16" × 16"	5½	4½	10
2	8" × 12"	5½	3½	9
2	8" × 12"	6	5½	11½
2	12" × 12"	6½	4½	11
2	8" × 12" 12" × 12"	6½	6	12½
2	12" × 12"	7	6½	13½
2	8" × 12"	7	5	12
3	8" × 12"	9	8	17
3	8" × 12"	7½	8½	16
3	8" × 12" 8" × 12" 12" × 12"	7½	7	14½
3	8" × 12" 12" × 12" 12" × 12"	8	7½	15½
3	12" × 12"	8½	8	16½
3	12" × 12"	9	10	19

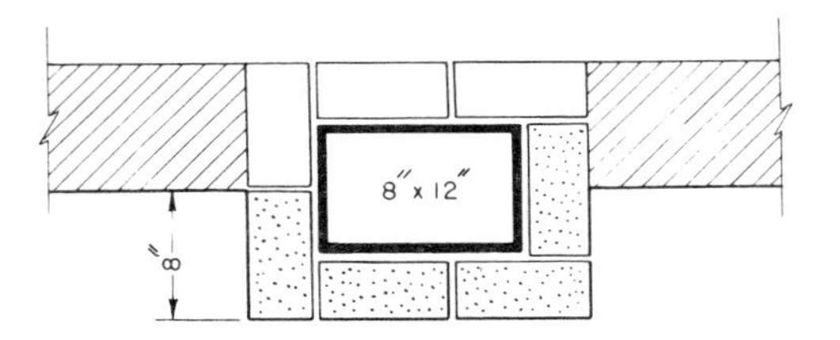

BRICK { 4 – FACE
3 – COMMON

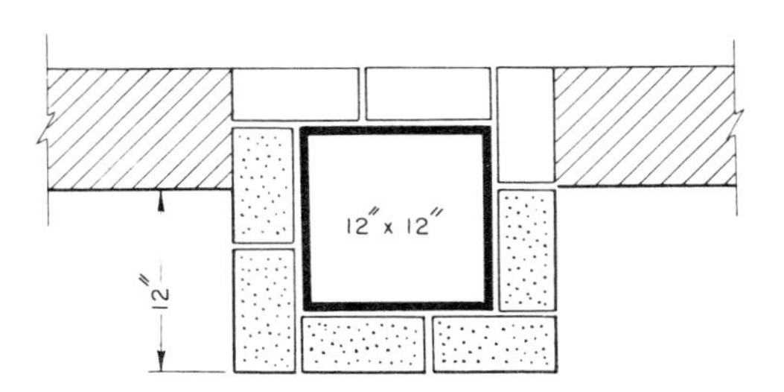

BRICK { 5 – FACE
3 – COMMON

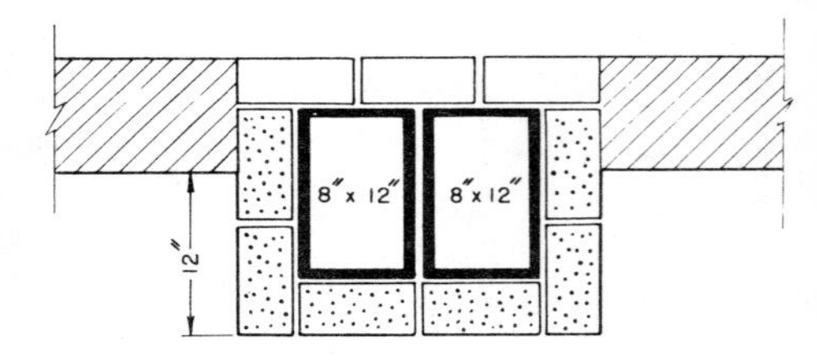
8" x 12"
8" x 12"
12"
BRICK 6 - FACE
3 - COMMON

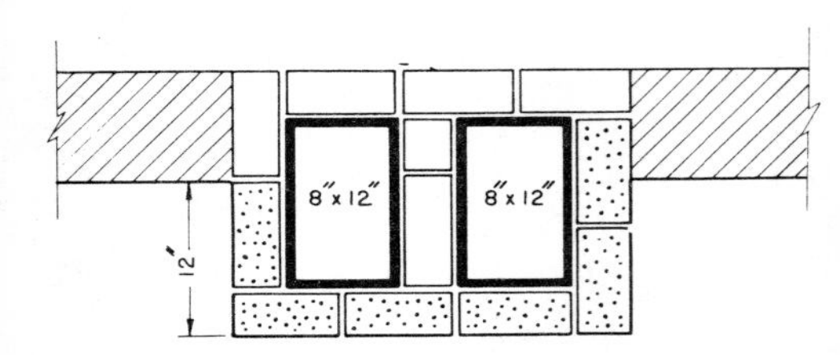
8" x 12"
8" x 12"
12"
BRICK 6 - FACE
5½ - COMMON

NUMBER OF FACE AND COMMON BRICK REQUIRED PER CROSS-SECTIONAL AREA OF CHIMNEYS
(Chimney extends 16" from wall.)

Number of Flues	Size of Flues	Face Brick	Common Brick	Total
1	4" × 8"	....	......	5
1	8" × 8"	....	......	6
1	8" × 12"	6	1	7
1	12" × 12"	6	2	8
1	12" × 16"	6½	2½	9
1	16" × 16"	6½	3½	10
2	8" × 12"	6½	2½	9
2	8" × 12"	7	4½	11½
2	12" × 12"	7½	3½	11
2	{ 8" × 12" 12" × 12"	7½	5	12½
2	12" × 12"	8	5½	13½
2	8" × 12"	8	4	12
3	8" × 12"	10	7	17
3	8" × 12"	8½	7½	16
3	{ 8" × 12" 8" × 12" 12" × 12"	8½	6	14½
3	{ 8" × 12" 12" × 12" 12" × 12"	9	6½	15½
3	12" × 12"	9½	7	16½
3	12" × 12"	10	9	19

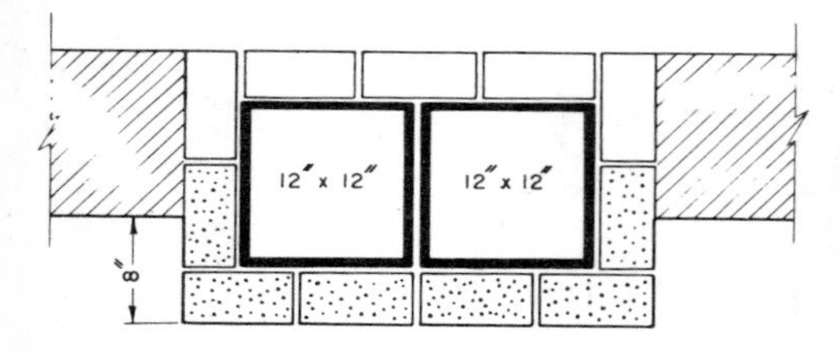

BRICK { 6 - FACE
5 - COMMON

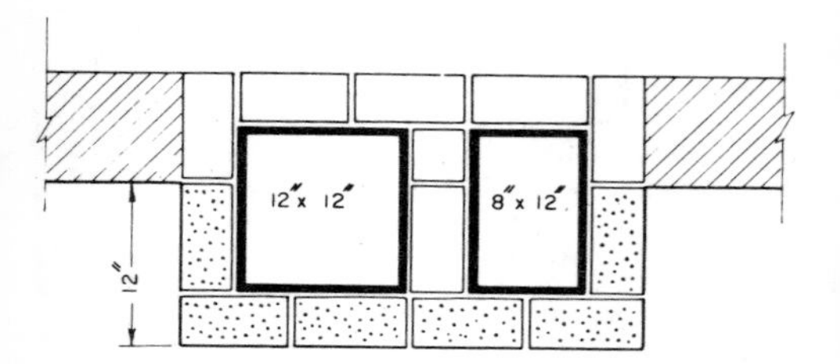

BRICK { 6 - FACE
6½ - COMMON

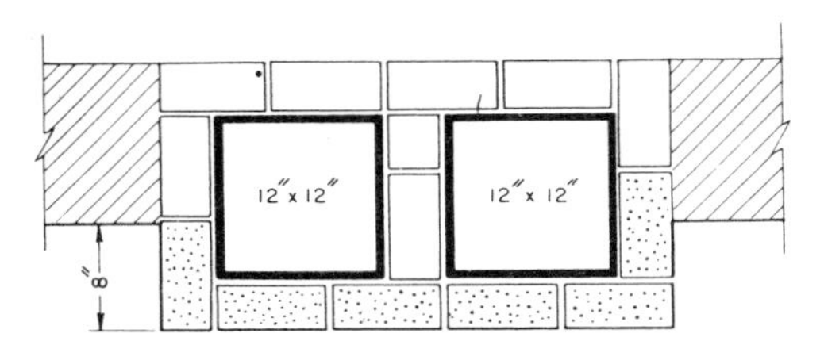

BRICK { 6 - FACE
7½ - COMMON

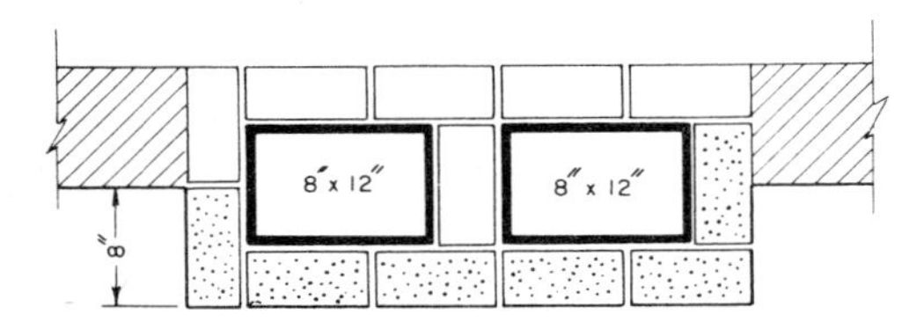

BRICK { 6 - FACE
6 - COMMON

NUMBER OF FACE AND COMMON BRICK REQUIRED PER CROSS-SECTIONAL AREA OF CHIMNEYS
(Chimney extends 20" from wall.)

Number of Flues	Size of Flue	Face Brick	Common Brick	Total
1	4" × 8"	....	......	5
1	8" × 8"	....	......	6
1	8" × 12"	....	......	7
1	12" × 12"	7	1	8
1	12" × 16"	7½	1½	9
1	16" × 16"	7½	2½	10
2	8" × 12"	7½	1½	9
2	8" × 12"	8	3½	11½
2	12" × 12"	8½	2½	11
2	8" × 12" 12" × 12"	8½	4	12½
2	12" × 12"	9	4½	13½
2	8" × 12"	9	3	12
3	8" × 12"	11	6	17
3	8" × 12"	9½	6½	16
3	8" × 12" 8" × 12" 12" × 12"	9½	5	14½
3	8" × 12" 12" × 12" 12" × 12"	10	5½	15½
3	12" × 12"	10½	6	16½
3	12" × 12"	11	8	19

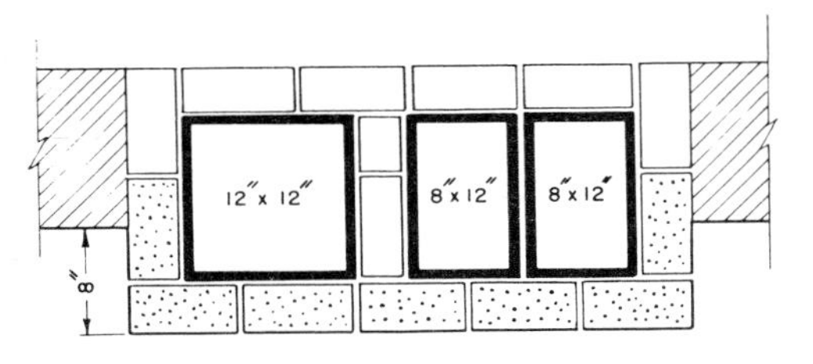

BRICK { 6½ – FACE
8 – COMMON

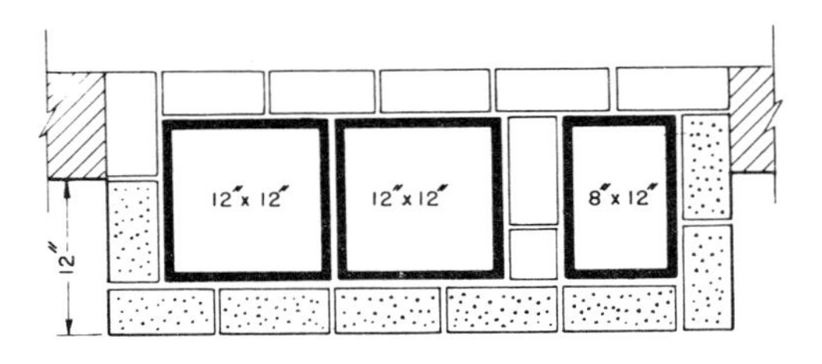

BRICK { 8 – FACE
7½ – COMMON

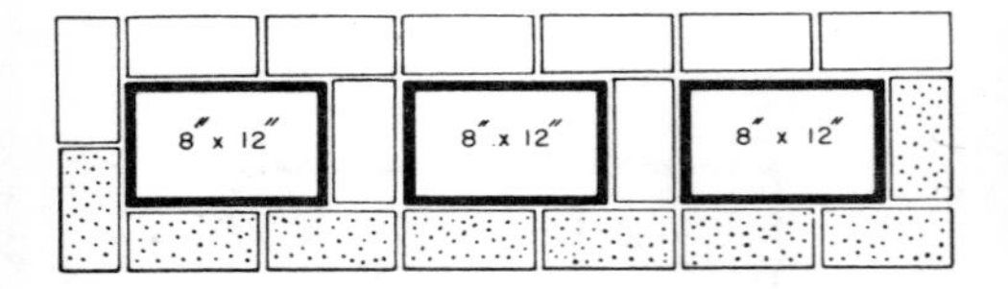

BRICK { 8 - FACE, 9 - COMMON }

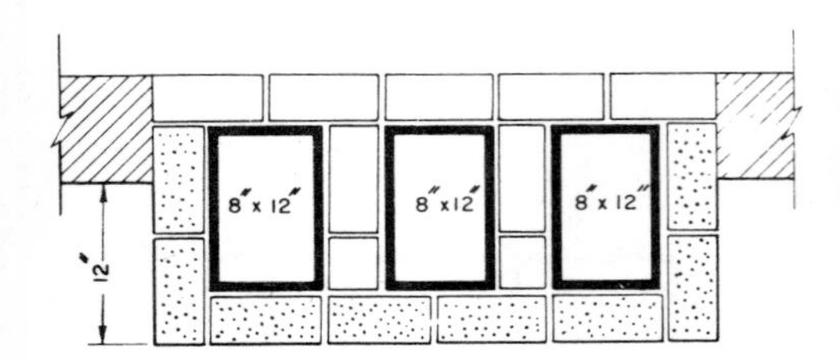

BRICK { 7½ - FACE, 8½ - COMMON }

EFFLORESCENCE ON BRICK WORK

Efflorescence is a light powder or crystallization deposited on the surface of brick work. It is caused by the evaporation of water carrying water soluble salts. Efflorescence is caused by soluble salts present in the wall material and moisture to carry these salts to the surface of the bricks. Good brick contain little soluble salts. Mortars are the main sources of the salts. Since moisture is necessary to carry soluble salts to the surface of the bricks, efflorescence is proof that there is faulty construction. Wet walls may be caused by defective flashing, gutters, and down-spouts, copings, poorly filled mortar joints.

MODULAR BRICK

Modular brick has become the standard for most government agencies. The better architectural offices also specify the use of modular brick design. Modular brick are easier to "take off" when a quantity survey is being prepared, are cheaper to lay, are coordinated with windows manufactured by members of the metal and wood institutes, and in their use eliminate cutting and fitting around openings, which represents a substantial saving on labor time.

On the basis of the 4" module taken both vertically and horizontally certain sizes of brick have been manufactured to fit within the 4" increments.

In the following illustration it can readily be seen that three 2⅔" modular bricks with a ½" mortar joint will fit into an 8" height, four 3" modular bricks will fit into a 12" height, and one 4" modular brick will fit into a 4" height.

A 2⅔" modular brick actually measures 2⅙" high, plus ½" mortar joint = 2⅔".

A 3" modular brick actually measures 2½" high, plus ½" mortar joint = 3".

A 4" modular brick actually measures 3½" high, plus ½" mortar joint = 4".

In the first of the following illustrations the modular brick measures actually 2⅙" × 3½" × 7½". If a ½" mortar joint is used, three courses of brick lay up perfectly to every two horizontal grid lines and one length of brick likewise fits two vertical grid lines. The same holds true of the other sizes of brick shown.

NOMINAL MODULAR SIZES OF BRICK
Nominal Sizes Include the Thickness of the Standard Mortar Joint

Thickness in Inches	Face Dimensions in Wall	
	Height, In.	Length, In.
4	2	12
4	2⅔	8
4	2⅔	12
4	4	8
4	4	12
4	5⅓	8
4	5⅓	12

NUMBER OF MODULAR BRICK FOR VARIOUS MORTAR JOINTS

Size of Brick Including Mortar Joint	Face Area of Brick (Sq. In.)	Number of Brick Per Sq. Ft.
4 × 2 × 12	24	4
4 × 2⅔ × 8	21.28	6.76
4 × 2⅔ × 12	31.92	4.51
4 × 4 × 8	32	4.5
4 × 4 × 12	48	3
4 × 5⅓ × 8	42.64	3.4
4 × 5⅓ × 12	63.96	2.26

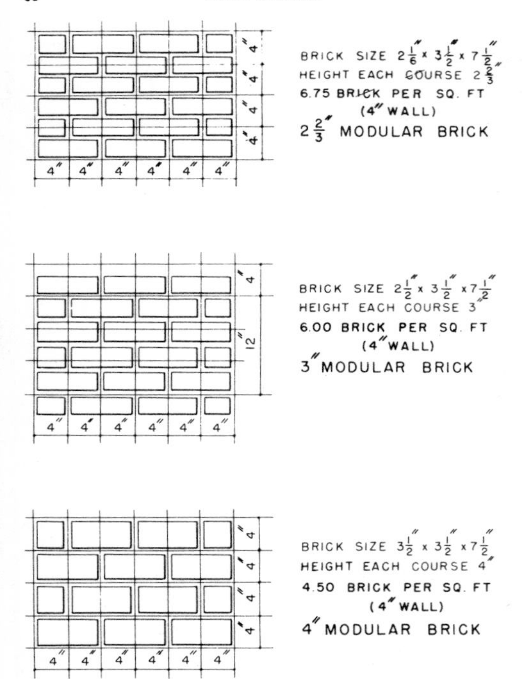

MORTAR

A mixture of cement, lime, and water. Use one part of cement, to three parts of sand to which is usually added some lime to make the mixture more workable.

A good brick mortar may consist of one part of cement, one part of lime and six parts of sand. Assuming the most commonly used brick joint of ½" thickness, for both end and bed joints, about 16½ cu. ft. of mortar are required for 1000 bricks.

Brick layers will lay between 500 and 600 bricks per 8 hr. day, and only about 300 or less face brick per 8 hr. day. About 600 sq. ft. of wall can be cleaned by the bricklayer in an 8 hr. day.

ESTIMATING QUANTITIES OF FLUE LINING

Most rectangular and round flue linings are manufactured in 2' lengths, although there are other sizes available.

In estimating the number of 2' sections required, subtract 4' from the total height of the chimney, and divide by 2.

For example: The chimney in the previous illustration (page 54) was 33' high. Since the flue lining begins at the point where the furnace connects to the chimney, a distance of approximately 4' high, the total height of the flue is therefore 33 − 4 = 29'.

29 ÷ 2 = 14.5 or 15 sections of 2' lengths

One of these lengths may be a special unit having an opening for the smoke-pipe connection.

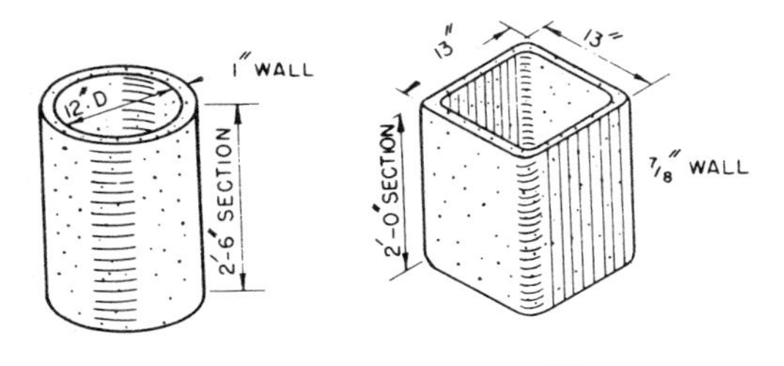

TYPICAL RESIDENTIAL FLUE LININGS
MADE FROM FIRE CLAY

DATA ON SIZES OF FLUES

Residential Flues—According to the National Board of Fire Underwriters flue linings must be of a certain size depending on the type of fuel and the system of heating.
For example: For solid or liquid fuels flue lining shall not be less than 70 sq. in. for warm air, hot water and low pressure steam heating appliances; for ordinary stoves, ranges and room heater not less than 40 sq. in.; and not less than 28 sq. in. for small special stoves and heaters; for fireplaces the flue area must be 50 sq. in. but at least 1/12th of the fireplace opening.

Non-Residential Flues—For non-residential flues and chimney sizes it is best to consult the Guide, published by the American Society of Heating and Ventilating Engineers, which gives formulae for determining size and height of chimneys when required draft and flue gas volume and temperature are known. Various boiler capacities require different flue and chimney sizes.

CROSS-SECTIONAL AREAS OF SMOKE FLUES FOR RESIDENCES*

Type of Heating Appliances Using Solid or Liquid Fuels	Cross-sectional Area, Sq. In. (Min.)
For warm-air, hot-water, and low-pressure steam heating appliances	70 sq. in.
Ordinary stoves, ranges, and room heaters	40 sq. in.
Small special stoves and heaters	28 sq. in.
For fireplaces	50 sq. in. (but not less than 1/12 of fireplace opening)

*According to the building code of the National Board of Fire Underwriters (1943 edition).

STANDARD SIZES OF MODULAR CLAY FLUE LININGS

Minimum Net Inside Area, Sq. In.	Nominal Dimensions,* In.	Outside Dimensions, In.	Minimum Wall Thickness, In.	Approximate Maximum Outside Radius, In.
15	4 × 8	3.5 × 7.5	.5	1
20	4 × 12	3.5 × 11.5	.625	1
27	4 × 16	3.5 × 15.5	.75	1
35	8 × 8	7.5 × 7.5	.625	2
57	8 × 12	7.5 × 11.5	.75	2
74	8 × 16	7.5 × 15.5	.875	2
87	12 × 12	11.5 × 11.5	.875	3
120	12 × 16	11.5 × 15.5	1	3
162	16 × 16	15.5 × 15.5	1.125	4
208	16 × 20	15.5 × 19.5	1.25	4
262	20 × 20	19.5 × 19.5	1.375	5
320	20 × 24	19.5 × 23.5	1.5	5
385	24 × 24	23.5 × 23.5	1.625	6

*Cross-section flue lining shall fit within rectangle of dimension corresponding to nominal size.

STANDARD FIRE-BRICK SIZES AND SHAPES*

Type	Size	Number Required per Sq. Ft.	
		Laid Flat	Laid on Edge
9" straight	9" × 4½" × 2½"	6.5	3.5
Small 9" brick	9" × 3½" × 2½"	6.5	4.5
Split brick	9" × 4½" × 1¼"	13.0	3.5
2" brick	9" × 4½" × 2"	8.0	3.5
Soap	9" × 2¼" × 2½"	6.5	7.0
Checker	9" × 2¾" × 2¾"	6.0	6.0

*Fire-brick prices will run from $75 to $125 per 1,000 brick.

TYPES OF CONCRETE BLOCKS

There are numerous concrete blocks on the market which are manufactured in two or three core units. The following are the types most commonly used:

1. Stretchers. Used for the wall proper.
2. Jamb units. Used around doors and windows.
3. Corner units. Used at outside corners.
4. Double corner units. Used for piers and pilasters.
5. Header units. Used for backing in masonry walls faced with bricks.

Most of the regular and special blocks are cast in half units to eliminate cutting on the job. Half units generally improve the appearance of concrete-block masonry construction.

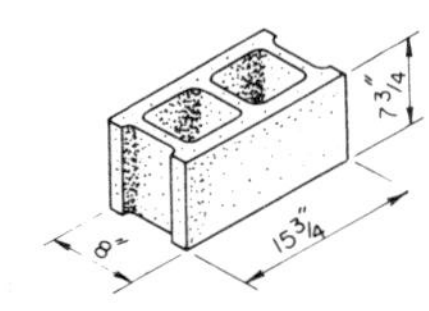

REGULAR STRETCHER UNIT

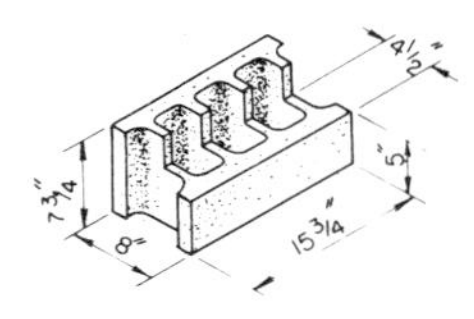

HEADER UNIT

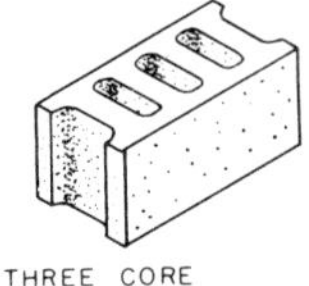

THREE CORE CONCRETE BLOCK

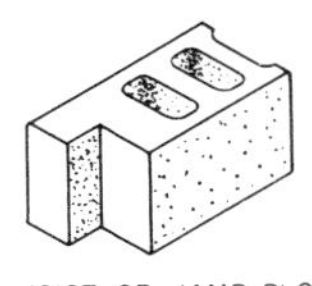

JOIST OR JAMB BLOCK

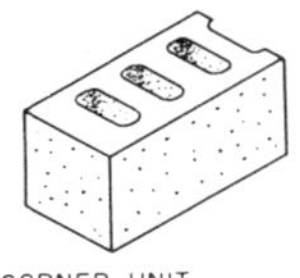

CORNER UNIT

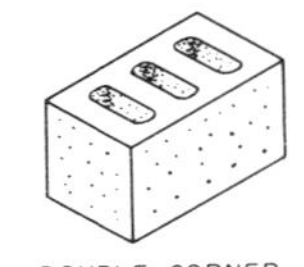

DOUBLE CORNER UNIT

TYPES OF CONCRETE BLOCKS

ESTIMATING CONCRETE BLOCKS

Estimate the number of square feet of wall of any thickness, and multiply by the number of blocks per 100 sq. ft. as given in the table on page 71. Deduct all areas for openings.

For example: Using an 8" × 12" × 16" block, on the plan of the basement 20' wide by 30' long and 8' high, the perimeter is

$$2 \times 20' + 2 \times 30' = 100'$$

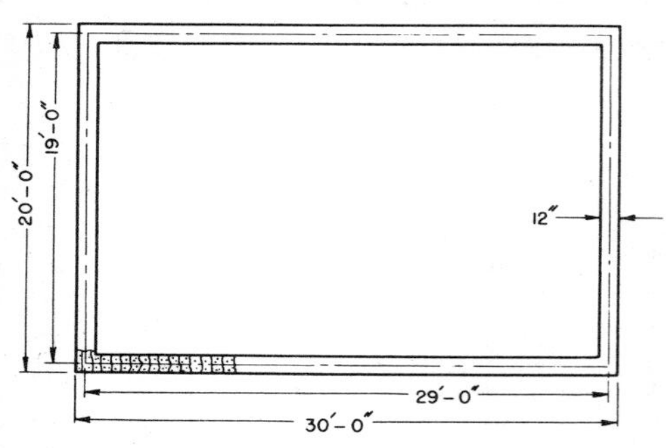

CONCRETE BLOCK
FOUNDATION WALL

From the perimeter deduct four times the thickness of the wall, or 4 × 12" = 48", or 4'.

Therefore, 100' − 4' = 96'

(Another method is to take the perimeter of a center line of the wall, or 2 × 29' + 2 × 19' = 96'.)

Multiply the perimeter by the height, or 8 × 96' = 768 sq. ft.

768 sq. ft. ÷ 100 = 7.68

There are 110 blocks in 100 sq. ft. (see table on concrete blocks per 100 sq. ft., page 71).

Therefore, 110 × 7.68 = 845 blocks

NUMBER OF CONCRETE-BLOCK COURSES

Divide the height of the wall by the height of the concrete block, thus:

8' wall, or 96" ÷ 8" = 12 courses

Each corner has 12 corner blocks.

4 corners = 4 × 12 = 48 corner blocks

Therefore, 845 blocks − 48 corner blocks = 797 standard blocks

If corner units are to be used adjacent to openings, divide the height of each opening by the nominal height of the blocks to be used, counting any fraction thus obtained as one additional. This is the probable number of full-sized units needed to construct the wall sections adjacent to the openings. Find the total number of special units required for all openings.

If concrete lintels are to be constructed over openings in the masonry walls, divide the length of each lintel by the length of the blocks used. In this manner find the total number of regular blocks to be replaced by lintels, and subtract from the total number of stretcher units previously calculated.

Note: Increase by 2% any number of blocks 100 or more to allow for waste and breakage. For any number less than 100 add two additional blocks for waste.

The amount of mortar required for the above condition is found in the following table; it is 3.25 cu. ft. per 100 sq. ft. of wall.

Therefore, $3.25 \times 7.68 = 25$ cu. ft.

CONCRETE BLOCKS AND MORTAR PER 100 SQ. FT.

Description, Size of Block, In.	Wall Thickness, In.	Weight per Unit, Lb.	Number of Units per 100 Sq. Ft. of Wall Area	Mortar Cu. Ft.	Weight, Lb. per 100 Sq. Ft. of Wall Area
8 × 8 × 16	8	50	110	3.25	5,850
8 × 8 × 12	8	38	146	3.50	6,000
8 × 12 × 16	12	85	110	3.25	9,700
8 × 3 × 16	3	20	110	2.75	2,600
9 × 3 × 18	3	26	87	2.50	2,500
12 × 3 × 12	3	23	100	2.50	2,550
8 × 3 × 12	3	15	146	3.50	2,550
8 × 4 × 16	4	28	110	3.25	3,450
9 × 4 × 18	4	35	87	3.25	3,350
12 × 4 × 12	4	31	100	3.25	3,450
8 × 4 × 12	4	21	146	4.00	3,500
8 × 6 × 16	6	42	110	3.25	5,000

HEIGHT OF CONCRETE BLOCK CONSTRUCTION BY COURSES FOR CONCRETE BLOCK 7½″ IN HEIGHT WITH ½″ AND ⅝″ MORTAR JOINTS

Height of Block	7½″	7⅝″	Height of Block	7½″	7⅝″
Joint Thickness	½″	⅝″	Joint Thickness	½″	⅝″
No. of Courses			No. of Courses		
1	8″		11	7′-4″	
2	1′-4″		12	8′-0″	
3	2′-0″		13	8′-8″	
4	2′-8″		14	9′-4″	
5	3′-4″		15	10′-0″	
6	4′-0″		16	10′-8″	
7	4′-8″		17	11′-4″	
8	5′-4″		18	12′-0″	
9	6′-0″		19	12′-8″	
10	6′-8″		20	13′-4″	

LOAD-BEARING WALL TILE

Find the areas of the walls following the formula given under Concrete Blocks (page 70).

Determine the thickness of the wall and the size of tile to be used. Deduct for all openings. Multiply the area by the number of tile required per square foot according to the following table.

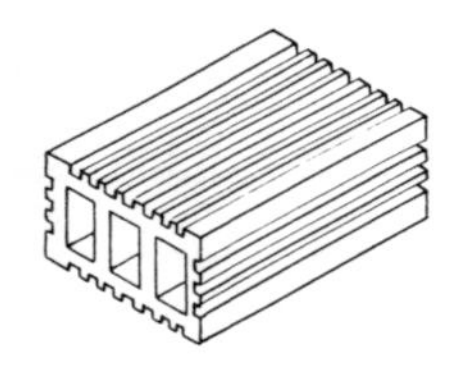

STRETCHER UNIT OF "BACK-UP" TILE

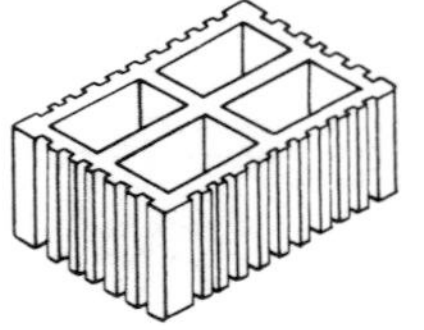

CLOSURE UNIT OF "BACK-UP" TILE

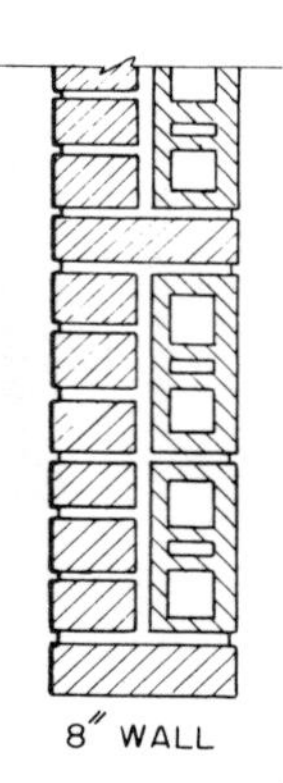

8″ WALL

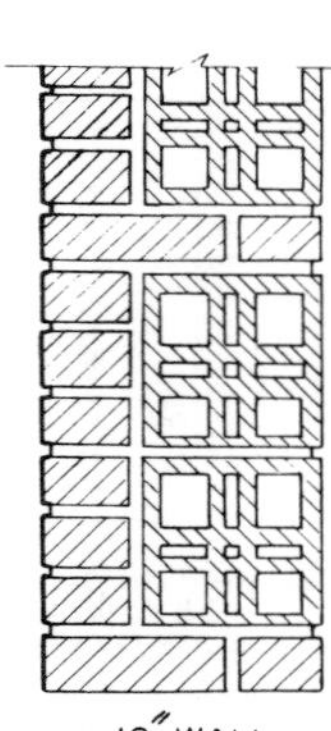

12″ WALL

BRICK AND TILE CONSTRUCTION

TILE, MORTAR, AND LABOR FOR LOAD-BEARING WALLS

Size of Tile, In.	Number per Sq. Ft.	Cu. Ft. Mortar per 100 Sq. Ft.	Hr. of Labor per 100 Sq. Ft.
3¾ × 12 × 12	1	2.25	4
6 × 12 × 12	1	3.25	4
8 × 12 × 12	1	4.15	5
12 × 12 × 12	1	5.75	5
SIDE CONSTRUCTION			
3¾ × 5 × 12	2.2	2.25	4
8 × 5 × 12	2.2	3.75	4
7¾ × 8 × 12	1.5	3.40	5
10¼ × 8 × 12	1.2	4.00	5

MATERIALS FOR MORTAR FOR 1,000 HOLLOW TILE
(1 part portland cement, 2 parts sand, 15% hydrated lime by volume)

Size of Tile, In.	Thickness of Wall, In.	Position Laid	Cu. Ft. of Mortar	Bags of Cement	Cu. Yd. of Sand	Bags of Lime
Partition tile:						
3 × 12 × 12	3	End or Side	18.7	6.2	.69	.75
4 × 12 × 12	4	End or side	22.2	7.4	.82	.89
6 × 12 × 12	6	End or side	22.2	7.4	.82	.89
Load-bearing tile:						
3¾ × 12 × 12	3¾	End or side	21.5	7.2	.80	.86
6 × 12 × 12	6	Side	30.0	10.0	1.11	1.20
6 × 12 × 12	6	End	26.0	8.7	.96	1.04
8 × 12 × 12	8	Side	37.6	12.5	1.40	1.50
8 × 12 × 12	8	End	33.2	11.1	1.23	1.33
10 × 12 × 12	10	Side	45.3	15.1	1.68	1.81
10 × 12 × 12	10	End	35.2	11.1	1.23	1.33
12 × 12 × 12	12	Side	52.9	17.6	1.96	2.11
12 × 12 × 12	12	End	37.1	12.4	1.38	1.49
3¾ × 5 × 12	3¾	Side	17.3	5.8	.64	.70
8 × 5 × 12	8	Side	33.4	11.1	1.24	1.33
8 × 12 × 12	8	End	33.6	11.2	1.24	1.34

GLASS BLOCKS

The following table shows the minimum sized openings needed for panels of various numbers of glass blocks, allowing for ¼" mortar joints. (The mortar consists of 1 part portland cement, 1 part hydrated lime, and 4 parts sand.)

For example: A panel of glass blocks, 5 blocks high and 7 blocks wide, using 7¾" × 7¾" blocks, will require an opening of 3'-4⅜" × 4'-8½".

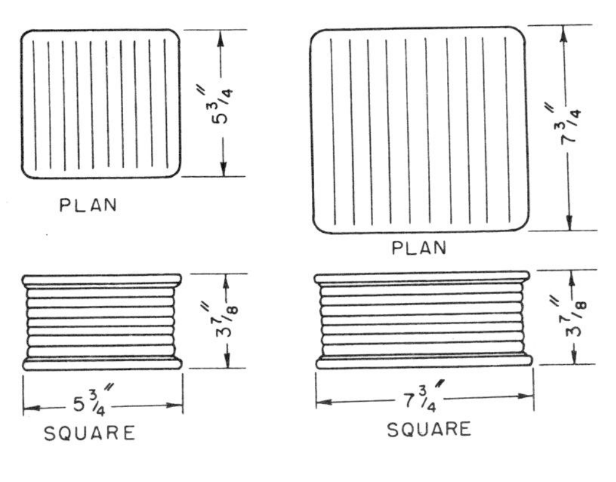

SIZES AND SHAPES OF GLASS BLOCKS

MINIMUM SIZED OPENINGS FOR GLASS-BLOCK PANELS IN MASONRY WALL

Number of Blocks	Block Size 5¾" × 5¾"		Block Size 7¾" × 7¾"		Block Size 11¾" × 11¾"	
	Height	Width	Height	Width	Height	Width
1	6⅜"	6½"	8⅜"	8½"	1'-0⅜"	1'-0½"
2	1'-0⅜"	1'-0½"	1'-4⅜"	1'-4½"	2'-0⅜"	2'-0½"
3	1'-6⅜"	1'-6½"	2'-0⅜"	2'-0½"	3'-0⅜"	3'-0½"
4	2'-0⅜"	2'-0½"	2'-8⅜"	2'-8½"	4'-0⅜"	4'-0½"
5	2'-6⅜"	2'-6½"	3'-4⅜"	3'-4½"	5'-0⅜"	5'-0½"
6	3'-0⅜"	3'-0½"	4'-0⅜"	4'-0½"	6'-0⅜"	6'-0½"
7	3'-6⅜"	3'-6½"	4'-8⅜"	4'-8½"	7'-0⅜"	7'-0½"
8	4'-0⅜"	4'-0½"	5'-4⅜"	5'-4½"	8'-0⅜"	8'-0½"
9	4'-6⅜"	4'-6½"	6'-0⅜"	6'-0½"	9'-0⅜"	9'-0½"
10	5'-0⅜"	5'-0½"	6'-8⅜"	6'-8½"	10'-0⅜"	10'-0½"

MINIMUM SIZED OPENINGS FOR GLASS-BLOCK PANEL IN FRAME OPENINGS

Number of Blocks	Block Size 12"		Block Size 8"	
	Width	Height	Width	Height
1	1'-1¾"	1'-1³⁄₁₆"	9¾"	9³⁄₁₆"
2	2'-1¾"	2'-1³⁄₁₆"	1'-5¾"	1'-5³⁄₁₆"
3	3'-1¹³⁄₁₆"	3'-1¼"	2'-1¾"	2'-1³⁄₁₆"
4	4'-1¹³⁄₁₆"	4'-1¼"	2'-9¾"	2'-9³⁄₁₆"
5	5'-1¹³⁄₁₆"	5'-1¼"	3'-5¾"	3'-5³⁄₁₆"
6	6'-1⅞"	6'-1⁵⁄₁₆"	4'-1¾"	4'-1³⁄₁₆"
7	7'-1⅞"	7'-1⁵⁄₁₆"	4'-9¾"	4'-9³⁄₁₆"
8	8'-1⅞"	8'-1⁵⁄₁₆"	5'-5¾"	5'-5³⁄₁₆"
9	9'-1¹⁵⁄₁₆"	9'-1⅜"	6'-1¾"	6'-1³⁄₁₆"
10	10'-2"	10'-1⅜"	6'-9¾"	6'-9³⁄₁₆"
11	....	....	7'-5¾"	7'-5³⁄₁₆"
12	....	....	8'-1¾"	8'-1³⁄₁₆"
13	....	....	8'-9¾"	8'-9³⁄₁₆"
14	....	....	9'-5¾"	9'-5³⁄₁₆"
15	....	....	10'-1¾"	10'-1³⁄₁₆"

ROUGH CARPENTRY

Rough carpentry includes all the framing and plain boarding of the house.

Lumber is the general expression for the material used in carpentry work. Pieces of large cross section are designated as timbers. Grades of lumber have been established by the trade association. Spruce, pine, and fir are used for framing lumber and sheathing, while oak, white pine, maple, and cypres are ordinarily used for flooring and finished work.

Boards are usually 1" thick and 2" to 10" wide. Boards less than 1" thick are figured as 1" thick for estimating purposes. . "Tongued and grooved" (T. & G.) sheathing boards are called 1 × 4, 1 × 6, and 1 × 8 but measure less than these dimensions.

Lumber for rough carpentry is figured according to the number of feet board measure. One foot board measure equals 144 sq. in. of wood surface 1" thick.

By writing the standard dimensions of lumber over 12' in the form of a fraction, the number of feet board measure of lumber is determined. For instance, 2 × 8" floor joist, 14' long, would equal

$$\frac{2 \times 8 \times 14}{12} = \frac{56}{3} = 18\tfrac{2}{3} \text{ b.f.m.}$$

In cases of several pieces the number of feet board measure in one piece is multiplied by the number of pieces involved.

When it is necessary to find the cost per foot board measure when the cost per 1,000 f.b.m. is known, divide the cost by 1,000. Similarly, when the cost per 100 f.b.m. is known, divide by 100 to get the cost per foot board measure.

Example: Find the cost of 40 pcs. of 2" × 4" × 18' at $150 per thousand.

SOLUTION:

$$\frac{40 \times 2 \times \cancel{4} \times \overset{6}{\cancel{18}}}{\underset{3}{\cancel{12}}} = 480 \text{ f.b.m.}$$

$150 ÷ 1,000 = $.15, the cost per f.b.m.
480 f.b.m. × .15 = $72, the cost of 480 f.b.m.

EXAMPLE: Find the cost of the following lumber, 10 pcs. of 2" × 4" × 10', 20 pcs. of 2" × 4" × 12', 25 pcs. of 2" × 4" × 14', when the cost of 2 × 4's is at $130.50 per thousand feet board measure.

SOLUTION:

$$\frac{10 \times \cancel{2} \times \overset{2}{\cancel{4}} \times 10}{\underset{3}{\cancel{12}}} = \frac{200}{3} = 66\tfrac{2}{3} \text{ or } 67 \text{ f.b.m.}$$

$$\frac{20 \times 2 \times 4 \times \cancel{12}}{\cancel{12}} = 160 \text{ f.b.m.}$$

$$\frac{25 \times 2 \times \cancel{4} \times 14}{\underset{3}{\cancel{12}}} = 233\tfrac{1}{3} \text{ or } 233 \text{ f.b.m.}$$

67 + 160 + 233 = 460, total number of f.b.m.
130.50 ÷ 1,000 = $.1305, cost per f.b.m.
460 × .1305 = $59.80, total cost

FIGURING THE NUMBER OF PIECES WITHIN A GIVEN NUMBER OF FEET BOARD MEASURE

When a certain job calls for a given number of feet board measure of lumber and the price has been determined, it will in many instances be necessary to find the number of certain sized pieces required.

EXAMPLE: Find the number of pieces of 2" × 6" × 20' which should be delivered for an order of 2,500 f.b.m.

SOLUTION:

$$\frac{2" \times \not{6}" \times 20}{\underset{2}{\not{12}}} = 20 \text{ f.b.m. in each piece}$$

2,500 ÷ 20 = 120.5 pcs. or 121 pcs. of 2" × 6" × 20'

FINDING THE NUMBER OF LINEAR FEET OF LUMBER FOR A GIVEN NUMBER OF FEET BOARD MEASURE

Smaller sized lumber is generally figured in linear feet. If an estimate calls for 135 feet board measure of 1" × 3" cross bridging and it is necessary to find the total linear feet, proceed as follows:

$$\frac{1 \times \not{3}}{\underset{4}{\not{12}}} = \tfrac{1}{4}, \text{ number of f.b.m. in each lin. ft. of material}$$

$$\frac{135}{4} = 33.75, \text{ or } 34, \text{ number of lin. ft. of material required}$$

STANDARD SIZES OF LUMBER SURFACED ON FOUR SIDES

Nominal Size	Thickness	Width
1" × 2"	25/32"	1 5/8"
1" × 3"	25/32"	2 5/8"
1" × 4"	25/32"	3 5/8"
1" × 5"	25/32"	4 5/8"
1" × 6"	25/32"	5 5/8"
1" × 8"	25/32"	7 1/2"
1" × 10"	25/32"	9 1/2"
1" × 12"	25/32"	11 1/2"
2" × 4"	1 5/8"	3 5/8"
2" × 6"	1 5/8"	5 5/8"
2" × 8"	1 5/8"	7 1/2"
2" × 10"	1 5/8"	9 1/2"
2" × 12"	1 5/8"	11 1/2"
4" × 6"	3 5/8"	5 1/2"
4" × 8"	3 5/8"	7 1/2"
4" × 10"	3 5/8"	9 1/2"
6" × 6"	5 1/2"	5 1/2"
6" × 8"	5 1/2"	7 1/2"
6" × 10"	5 1/2"	9 1/2"
8" × 8"	7 1/2"	7 1/2"
8" × 10"	7 1/2"	9 1/2"
8" × 12"	7 1/2"	11 1/2"

RAPID LUMBER COMPUTATIONS IN FOOT BOARD MEASURE FOR GIVEN LENGTHS

Size, In.	Length, Ft.							
	8'	10'	12'	14'	16'	18'	20'	22'
1" × 2"	1⅓	1⅔	2	2⅓	2⅔	3	3⅓	3⅔
1" × 3"	2	2½	3	3½	4	4½	5	5½
1" × 4"	2⅔	3⅓	4	4⅔	5⅓	6	6⅔	7⅓
1" × 5"	3⅓	4⅙	5	5⅝	6⅔	7½	8⅓	9⅙
1" × 6"	4	5	6	7	8	9	10	11
1" × 8"	5⅓	6⅔	8	9⅓	10⅔	12	13⅓	14⅔
1" × 10"	6⅔	8⅓	10	11⅔	13⅓	15	16⅔	18⅓
1" × 12"	8	10	12	14	16	18	20	22
2" × 4"	5⅓	6⅔	8	9⅓	10⅔	12	13⅓	14⅔
2" × 6"	8	10	12	14	16	18	20	22
2" × 8"	10⅔	13⅓	16	18⅔	21⅓	24	26⅔	29⅓
2" × 10"	13⅓	16⅔	20	23⅓	26⅔	30	33⅓	36⅔
2" × 12"	16	20	24	28	32	36	40	44
4" × 4"	10⅔	13⅓	16	18⅔	21⅓	24	26⅔	29⅓
4" × 6"	16	20	24	28	32	36	40	44
4" × 8"	21⅓	26⅔	32	37⅓	42⅔	48	53⅓	58⅔
4" × 10"	26⅔	33⅓	40	46⅔	53½	60	66⅔	73⅓
6" × 6"	34	30	36	42	48	54	60	66
6" × 8"	32	40	48	56	64	72	80	88
6" × 10"	40	50	60	70	80	90	100	110
8" × 8"	42⅔	53⅓	64	74⅔	85⅓	96	106⅔	117⅓
8" × 10"	53⅓	66⅔	80	93⅓	106⅔	120	133⅓	146⅔
8" × 12"	64	80	96	112	128	144	160	176

RAPID LUMBER COMPUTATIONS IN FOOT BOARD MEASURE

1 × 3	Divide lin. ft. by 4
1 × 4	Divide lin. ft. by 3
1 × 6	Divide lin. ft. by 2
1 × 8	Multiply lin. ft. by 2, and divide by 3
1 × 10	Multiply lin. ft. by 10, and divide by 12
1 × 12	Lin. ft. and f.b.m. the same
2 × 3	Divide lin. ft. by 2
2 × 4	Multiply lin. ft. by 2, and divide by 3
2 × 8	Add to lin. ft. ⅓ of amount
2 × 10	Multiply lin. ft. by 10, and divide by 6
2 × 12	Multiply lin. ft. by 2
3 × 3	Multiply lin. ft. by 3, and divide by 4
3 × 4	Lin. ft. and f.b.m. the same
3 × 6	Add to lin. ft. ½ the amount
3 × 8	Multiply lin. ft. by 2
3 × 10	Multiply lin. ft. by 10, and divide by 4
3 × 12	Multiply lin. ft. by 3
4 × 4	Add to lin. ft. ⅓ of amount
4 × 6	Multiply lin. ft. by 2
4 × 8	Multiply lin. ft. by 3, and subtract ⅓ lin. ft. from amount
4 × 10	Multiply lin. ft. by 10, and divide by 3
4 × 12	Multiply lin. ft. by 4
8 × 8	Multiply lin. ft. by 5⅓
10 × 10	Multiply lin. ft. by 100, and divide by 12
12 × 12	Multiply lin. ft. by 12
14 × 14	Multiply lin. ft. by 16⅓

RAPID LUMBER COMPUTATIONS IN FOOT BOARD MEASURE
(Continued)

16 × 16	Multiply lin. ft. by 21⅓
18 × 18	Multiply lin. ft. by 27
20 × 20	Multiply lin. ft. by 100, and divide by 3
22 × 22	Multiply lin. ft. by 40⅓
24 × 24	Multiply lin. ft. by 48

ESTIMATING EXTERIOR AND INTERIOR WALL STUDDING

The number of studding in each wall section can be found by dividing the length of the section by the center-to-center distance between the studding, counting any fraction as one additional studding.

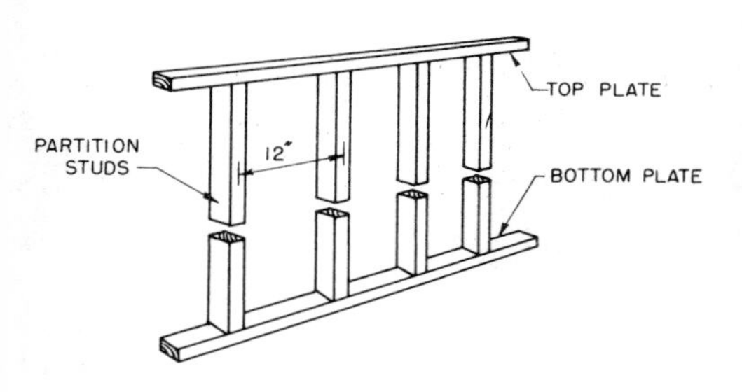

For inside partitions, add one additional studding to the number found.

For exterior wall construction with only a single studding for a corner post, add one additional studding to the number previously found.

For exterior wall construction in which corner posts consist of two studding, add three additional studding to the number previously found.

In estimating the number of studding in any wall adjacent to two other walls for which lumber for corner posts has already been included, subtract one from the number of studding previously found.

EXAMPLE: Find the number of 2" × 4" studding required for the outside walls and inside partition walls of the lakeside cottage shown on the following page. Assume a studding spacing of 16" on center and a studding height of 8'-0".

SOLUTION: The length of the outside walls and inside partitions as determined by scale measurements from the top to the bottom of the floor plan are as follows:

25'-8" + 6'-3" + 2'-10" + 25'-8" + 5'-0" = 65'-5", combined lengths of walls taken horizontally.

The lengths of the outside walls and inside walls taken vertically and from left to right are as follows:

26'-0" + 2'-6" + 3'-6" + 12'-6" + 12'-6" + 26'-0" = 83'-0", combined lengths of these walls.

65'-5" + 83'-0" = 148'-5" or 149', total length of combined walls.

Considerable additional material is required for constructing corner posts and framing around openings for doors and windows. An allowance of one studding per foot is commonly made in estimating the lumber required to construct these wall frames. Therefore, the number of linear feet of outside wall and inside partitions is the number of studding needed.

The total linear length is 149', which is also the number of pieces of 2" × 4" × 8' studding needed.

If the outside walls require a different length of studding from the inside partitions, the number of studding for the outside walls and inside partitions must be found separately.

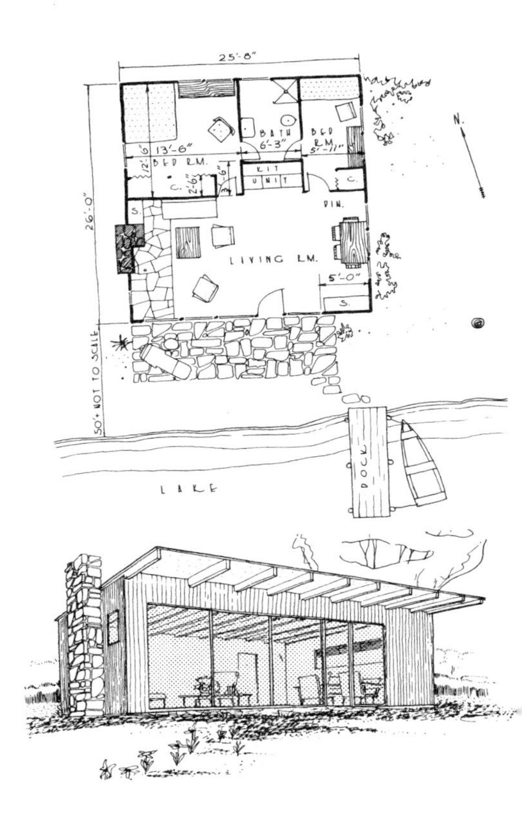
25'-8"
26'-0"
BATH
6'-3"
BED
RM.
5'-11"
13'-6"
12'-6"
BED RM.
C.
2'-6"
6"
3'
KIT
UNIT
C.
DIN.
S.
LIVING RM.
5'-0"
S.
N.
50'+ NOT TO SCALE
DOCK
LAKE

PARTITION STUDS, NUMBER REQUIRED FOR ANY SPACING

Distance Studs Are Apart			
12"	16"	20"	24"
Multiply length of partition by factor, and add 1 stud.			
1.0	.75	.60	.50

Add Additional for top and bottom plates.

B.F. OF LUMBER FOR STUD PARTITIONS
2" × 4" Studs 16" O.C. with Single Bottom and Double Top Plate

Length of Partition Wall in Feet	Height of Stud Partition					
	7½ Ft.	8 Ft.	8½ Ft.	9 Ft.	9½ Ft.	10 Ft.
5	30	31	33	34	35	37
6	37	39	40	42	44	45
7	44	46	48	50	52	54
8	46	48	50	52	54	56
9	53	55	58	60	62	65
10	60	63	65	68	71	73
11	67	70	73	76	79	82
12	69	72	75	78	81	84
13	76	79	83	86	89	93
14	83	87	90	94	98	101
15	90	94	98	102	106	110
16	92	96	100	104	108	112
17	99	103	108	112	116	121
18	106	111	115	120	125	129
19	113	118	123	128	133	138
20	115	120	125	130	135	140
21	122	127	133	138	143	149
22	129	135	140	146	152	157
23	136	142	148	154	160	166
24	138	144	150	156	162	168
25	145	151	158	164	170	177
26	152	159	165	172	179	185
27	159	166	173	180	187	194

NUMBER OF WOOD JOISTS REQUIRED FOR VARIOUS FLOOR LENGTHS AND SPACINGS

Joist Spacing Inches	Distance in Feet Covered by Joists—Joists are at Right Angles to Distance																		
	10	11	12	13	14	15	16	17	18	19	20	21	22	23	24	25	26	27	28
16	9	10	10	11	12	12	13	14	15	16	16	17	18	18	19	20	21	22	22
18	8	9	9	10	11	11	12	13	13	14	15	15	16	16	17	17	18	19	20
20	7	8	9	9	10	10	11	12	12	13	13	14	15	15	16	16	17	18	18
24	6	7	7	8	8	9	9	10	10	11	11	12	12	13	13	14	14	15	15
26	6	6	7	7	8	8	9	9	10	10	11	11	12	12	13	13	14	14	14
28	6	6	7	7	7	8	8	9	9	10	10	10	11	11	12	12	13	13	13
30	5	6	6	7	7	7	8	8	9	9	9	10	10	11	11	11	12	12	12
32	5	6	6	6	7	7	7	8	8	9	9	9	10	10	10	11	11	12	12
34	5	5	6	6	6	7	7	7	8	8	8	9	9	9	10	10	11	11	11
36	5	5	5	6	6	6	7	7	7	8	8	8	9	9	9	10	10	10	11

Example: A distance of 18 ft. is covered by floor joists 16" on center. Therefor, $18 \times 12 = 216$ inches. $216 \div 16 = 13.5$ joists, or 14 joists plus $1 = 15$.

NOTE: A distance of 3 ft. at 12" spacing requires 4 joists.
A distance of 4 ft. at 12" spacing requires 5 joists.
Therefor; one joist has been added to every distance.

ESTIMATING THE NUMBER OF WOOD JOISTS

To find the number of joists required over a certain area, divide the joist spacing into the length or span to be covered by such joists, and add one joist.

EXAMPLE: An area 20'-0" × 40'-0" is to be covered by floor joists spaced 16" on center. How many joists are required?

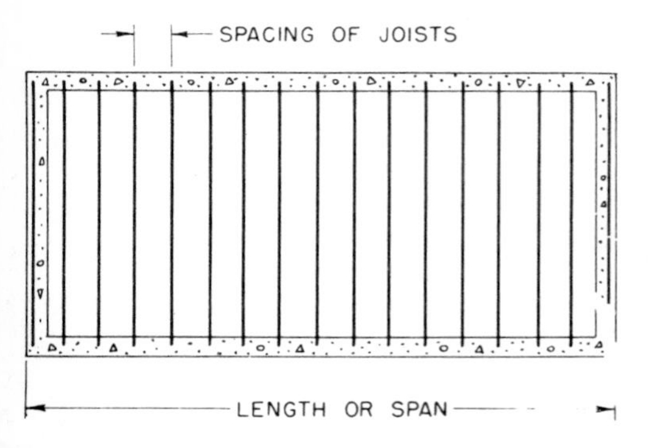

SOLUTION:

40" × 12" = 480", length of span, in.
480 ÷ 16 = 30, number of joists.
30 + 1 = total number of joists including extra joist at end of span

The following table gives the number of wood joists for spans ranging from 6'-0" to 40'-0" and a joist spacing from 12" to 60". The one extra joist required at the end is included in the table.

Wherever partitions run parallel to the joists, allow for doubling of joists.

MAXIMUM SPANS FOR FLOOR JOISTS NO. 1 COMMON

Live Load for Residential Use of 40 lb. per Sq. Ft. Uniformly Distributed with and without Plastered Ceiling

American Standard Lumber Sizes		Dist. on Cntr.	Maximum Clear Span between Supports					
			So. Pine and Douglas Fir		Western Hemlock		Spruce	
Nominal	Net		Unplstd.	Plstd.	Unplstd.	Plsta.	Unplstd.	plstd.
2" × 6"	1⅝" × 5⅝"	12"	12'-0"	10'-0"	11'-6"	9'-6"	10'-11"	9'-1"
		16"	10'-6"	9'-1"	10'-0"	8'-8"	9'-6"	8'-3"
3" × 6"	2⅝" × 5⅝"	12"	15'-0"	11'-8"	14'-4"	11'-2"	13'-8"	10'-6"
		16"	13'-1"	10'-8"	12'-6"	10'-2"	12'-0"	9'-8"
2" × 8"	1⅝" × 7½"	12"	15'-11"	13'-3"	15'-3"	12'-8"	14'-6"	12'-0"
		16"	13'-11"	12'-1"	13'-4"	11'-7"	12'-8"	11'-0"
3" × 8"	2⅝" × 7½"	12"	19'-8"	15'-4"	18'-10"	14'-8"	17'-11"	13'-11"
		16"	17'-4"	14'-0"	16'-7"	13'-5"	15'-9"	12'-9"
2" × 10"	1⅝" × 9½"	12"	19'-11"	16'-8"	19'-1"	16'-0"	18'-3"	15'-2"
		16"	17'-4"	15'-3"	16'-8"	14'-7"	15'-11"	13'-10"
3" × 10"	2⅝" × 9½"	12"	24'-7"	19'-3"	23'-6"	18'-5"	22'-5"	17'-6"
		16"	21'-8"	17'-8"	20'-9"	16'-11"	19'-9"	16'-1"
2" × 12"	1⅝" × 11½"	12"	23'-11"	20'-1"	22'-11"	19'-3"	21'-10"	18'-3"
		16"	20'-11"	18'-5"	20'-1"	17'-7"	19'-3"	16'-9"
3" × 12"	2⅝" × 11½"	12"	29'-4"	23'-1"	28'-1"	22'-1"	25'-5"	20'-11"
		16"	25'-11"	21'-3"	24'-10"	20'-4"	22'-5"	19'-4"

MAXIMUM SPANS FOR FLOOR JOISTS NO. 1 COMMON *(Continued)*

American Standard Lumber Sizes		Dist. on Cntr.	Maximum Clear Span between Supports					
			So. Pine and Douglas Fir		Western Hemlock		Spruce	
Nominal	Net		Unplstd.	Plstd.	Unplstd.	Plstd.	Unplstd.	Plstd.
2"×14"	1⅝"×13½"	12"	27'-8"	23'-5"	26'-6"	22'-6"	25'-3"	21'-2"
		16"	24'-4"	21'-5"	23'-4"	20'-6"	22'-3"	19'-6"
3"×14"	2⅝"×13½"	12"	....	26'-11"	....	25'-9"	30'-0"	24'-5"
		16"	....	24'-10"	....	23'-9"	27'-6"	22'-6"
Live Load of 60 lb. per Sq. Ft. Uniformly Distributed with and without Plastered Ceiling								
3"×6"	2⅝"×5⅝"	12"	12'-7"	10'-6"	12'-1"	10'-1"	11'-6"	9'-7"
		16"	11'-0"	9'-7"	10'-6"	9'-3"	10'-0"	8'-9"
2"×8"	1⅝"×7½"	12"	13'-4"	12'-0"	12'-9"	11'-6"	12'-2"	10'-11"
		16"	11'-8"	11'-8"	11'-2"	10'-6"	10'-8"	9'-11"
3"×8"	2⅝"×7½"	12"	16'-8"	14'-0"	16'-0"	13'-4"	15'-2"	12'-9"
		16"	14'-6"	12'-10"	14'-0"	12'-3"	13'-4"	11'-7"
2"×10"	1⅝"×9½"	12"	16'-10"	15'-2"	16'-1"	14'-6"	15'-4"	13'-9"
		16"	14'-8"	13'-10"	14'-0"	13'-3"	13'-4"	12'-7"
4"×8"	3⅝"×7½"	12"	19'-10"	15'-5"	18'-11"	14'-10"	18'-1"	14'-0"
		16"	17'-4"	14'-1"	16'-8"	13'-6"	15'-11"	12'-11"
3"×10"	2⅝"×9½"	12"	20'-11"	17'-7"	20'-0"	16'-10"	19'-1"	15'-11"
		16"	18'-4"	16'-1"	17'-6"	15'-5"	16'-9"	14'-7"
2"×12"	1⅝"×11½"	12"	20'-2"	18'-3"	19'-4"	17'-6"	18'-5"	16'-8"
		16"	17'-8"	16'-8"	16'-11"	16'-0"	16'-2"	15'-2"
3"×12"	2⅝"×11½"	12"	25'-0"	21'-1"	24'-0"	20'-3"	22'-10"	19'-3"
		16"	22'-0"	19'-4"	21'-1"	18'-6"	20'-1"	17'-7"
3"×14"	2⅝"×13½"	12"	29'-1"	24'-7"	27'-10"	33'-7"	26'-7"	22'-5"
		16"	25'-8"	22'-7"	24'-6"	21'-7"	23'-5"	20'-6"
Live Load of 75 lb. per Sq. Ft. Uniformly Distributed with and without Plastered Ceiling								
2"×8"	1⅝"×7½"	12"	12'-0"	11'-4"	11'-6"	10'-10"	11'-0"	10'-3"
		16"	10'-6"	10'-4"	10'-0"	9'-10"	9'-7"	9'-5"
3"×8"	2⅝"×7½"	12"	15'-2"	13'-2"	14'-6"	12'-8"	13'-10"	12'-1"
		16"	12'-2"	12'-1"	12'-8"	11'-7"	12'-5"	11'-0"
2"×10"	1⅝"×9½"	12"	15'-3"	14'-3"	14'-7"	13'-8"	13'-11"	12'-11"
		16"	13'-3"	13'-0"	12'-8"	12'-5"	12'-1"	11'-10"
4"×8"	3⅝"×7½"	12"	17'-7"	14'-7"	16'-11"	14'-0"	16'-2"	13'-3"
		16"	15'-5"	13'-4"	14'-10"	12'-9"	14'-1"	12'-2"
3"×10"	2⅝"×9½"	12"	19'-0"	16'-8"	18'-3"	15'-11"	17'-5"	15'-2"
		16"	16'-8"	15'-2"	15'-11"	14'-7"	15'-3"	13'-9"
2"×12"	1⅝"×11½"	12"	18'-4"	17'-2"	17'-6"	16'-5"	16'-9"	15'-8"
		16"	16'-0"	15'-9"	15'-2"	15'-1"	14'-7"	14'-3"
4"×10"	3⅝"×9½"	12"	22'-1"	18'-4"	21'-2"	17'-7"	20'-1"	16'-8"
		16"	18'-5"	16'-10"	18'-7"	16'-2"	17'-9"	15'-3"
3"×12"	2⅝"×11½"	12"	22'-10"	20'-1"	21'-11"	19'-2"	20'-8"	18'-2"
		16"	20'-1"	18'-4"	19'-3"	17'-7"	18'-4"	16'-8"
3"×14"	2⅝"×13½"	12"	26'-7"	23'-5"	25'-3"	22'-3"	24'-3"	21'-3"
		16"	23'-4"	21'-6"	22'-4"	20'-6"	21'-4"	19'-6"

Note: Deflection limited to 1/360 of the span. Dead load figured to include weight of joists, lath, and plaster ceiling (10 lb.) and sub floor and finish floor. Data supplied by National Lumber Manufacturers Association.

ESTIMATING QUANTITIES OF BRIDGING

Bridging should be placed in all spans of joists 8' or longer. In long spans of joists bridging should be placed at intervals of approximately 6 to 8'. Bridging material is generally 1" × 3", 1" × 4", and 2" × 2".

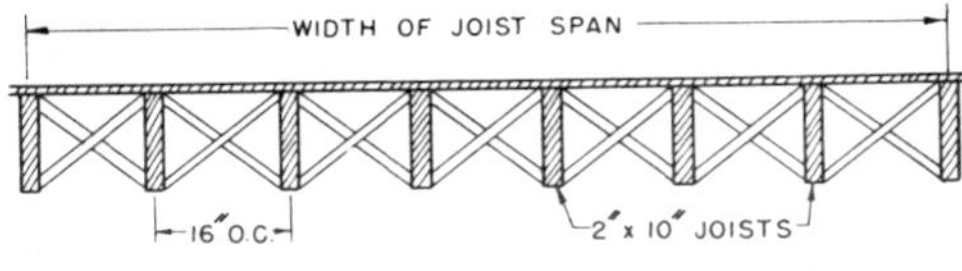

ESTIMATING QUANTITIES OF BRIDGING

The amounts of material required can be found from the number of spaces between the joists and the number of linear feet of material required for each space between the joists. The following table will give the linear feet of bridging required between joists for various spacings and joist sizes.

EXAMPLE: Find the linear feet of 1" × 4" cross bridging between joists spaced 16" on center covering a width of 20'-0".

SOLUTION:

9' = 108"
108 ÷ 16 = 6.75 or 7, number of spaces to be bridged in each joist section

From the following table, 16" on center and 2" × 10" joist size, there are 3 lin. ft. of bridging required between each joist space. A 9' length of 1" × 4" cross bridging will be a suitable stud length.

9 ÷ 3 = 3, the number of sets of bridging which can be cut from each piece.

7 ÷ 3 = 2.33 or 3, the number of pieces of 1" × 4" × 9' required.

NUMBER OF LINEAR FEET OF BRIDGING REQUIRED FOR EACH SPACE BETWEEN THE JOISTS

Size of Joists	Center-to-center Distance between Joists			
	12"	16"	20"	24"
2" × 6"	2	3	3½	4
2" × 8"	2½	3	3½	4
2" × 10"	2½	3	3½	4½
2" × 12"	3	3½	4	4½
2" × 14"	3	3½	4	4½

NUMBER OF PIECES OF DOUBLE BRIDGING PER 100 SQ. FT. OF FLOOR

Joists Up to 12' Long			
12" Centers		16" Centers	
Number of Pcs.	Lin. Ft.	Number of Pcs.	Lin. Ft.
20	30	10	24
Joists Up to 20' Long			
40	60	32	48

STEEL BRIDGING FOR WOOD JOISTS

Steel bridging is usually made in one piece and is easily bent to make double bridging. It requires only one nail at each end or four nails per piece and is furnished for 8", 10", and 12" joists and 12" and 16" spacing.

COMMON RAFTERS

The span of a roof is the distance between the outer edges of plates supporting the rafters, commonly the width of the building in most simple structures. The ridge of the roof is the part at the highest point or level at which the opposite pairs of rafters meet. The rise of the roof is the height of the roof above the level of the plates supporting the rafters. The distance which a part of a roof extends beyond the outside walls is called the overhang of the roof. The run of a rafter is the horizontal distance within the building directly underneath the rafter.

The slope of the roof is expressed in terms of rise of the roof in inches for each foot of run of rafters. Another common way of expressing the slope of roofs is to express it in terms of pitch. Pitch is the ratio of the rise of the roof to the span. A pitch of ½ means that the rise is one-half of the span of the roof.

The pitch of roofs can be changed to the equivalent rise of the roof in inches per foot, and the rise can be changed to an equivalent pitch as follows:

1. To find the rise of a roof per foot from the pitch, multiply the pitch by 24.
2. To find the pitch of a roof from the rise of a roof per foot, divide the rise by 24.

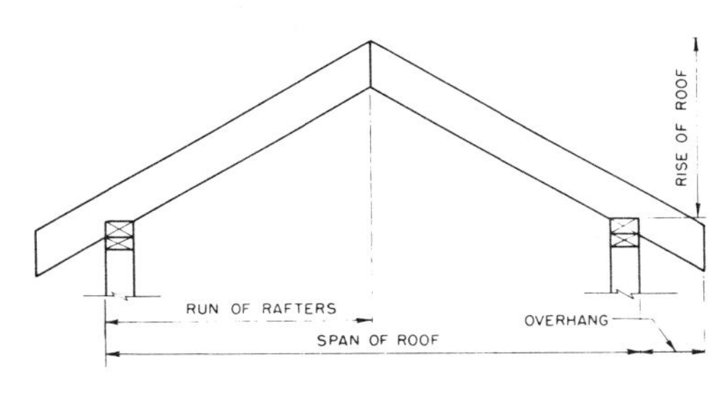

COMMON RAFTERS

LENGTH OF RAFTERS ON THE MEASURING LINE

The length of a common, hip, or valley rafter along the measuring line is the distance from the outer edge of the plate to the center of the ridge or the ridge board, measured parallel to the sides of the rafters.

The lengths along the measuring lines of all rafters on roofs of equal pitch can be determined from data given in the following table, which gives the pitches from $\frac{1}{12}$ to $\frac{3}{4}$. The rise of the roof per foot is also given in the table for each pitch.

To find the exact length of the rafter from the outer edge of the plate to the face of the ridge board, proceed as follows: Assume a roof span of 30'-0" with a roof pitch of $\frac{1}{4}$ and a ridge board 1" thick.

1. $\frac{1}{2}$ of 30' = 15', run of rafter.
2. $15 \times 13.42 = 201.30$", length of common rafter from plate to center of ridge board.
3. 201" = 16'-8".
   $\frac{.30}{1} \times \frac{8}{8} = \frac{2.40}{8} = .30$ or $\frac{3}{8}$" (to nearest $\frac{1}{8}$").
   201.30" = 16'-8$\frac{3}{8}$", length from plate to ridge expressed in dimension form.
4. 16'-8$\frac{3}{8}$" $- \frac{13}{16}$ ($\frac{1}{2}$ thickness of ridge board 2×8 or $1\frac{5}{8} \times 7\frac{1}{2}$ actual size. $\frac{1}{2}$ of $1\frac{5}{8} = \frac{13}{16}$.

Therefore, 16'-8$\frac{3}{8}$" – $\frac{13}{16}$" = 16'-7$\frac{9}{16}$"
the exact length of the common rafter from the plate to the ridge board.

The table on page 87 on Allowances in Laying Out Rafters will give the amount to be subtracted from the measuring line when a ridge board of 1" in thickness is used.

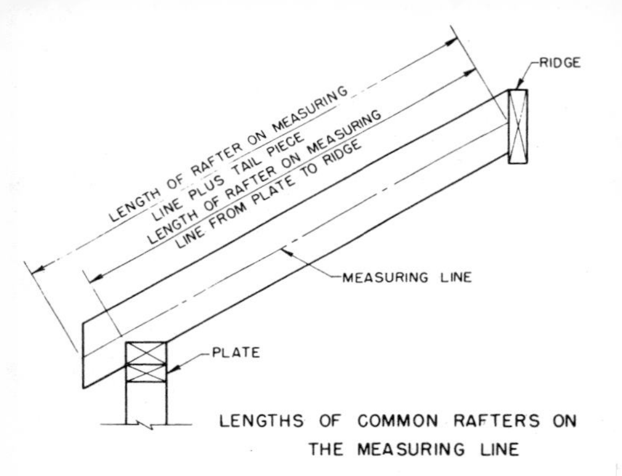

LENGTHS OF COMMON RAFTERS ON THE MEASURING LINE

LENGTHS OF COMMON RAFTERS (ALONG MEASURING LINE) OF EQUAL PITCH

Pitch of Roof	Length of Common Rafter per Ft. of Run	Length of Hip or Valley Rafter per Ft. of Run of Common Rafter	Rise of Roof per Ft.
1/12	12.17"	17.09"	2"
1/8	12.37"	17.23"	3"
1/6	12.65"	17.44"	4"
5/24	13.00"	17.69"	5"
1/4	13.42"	18.00"	6"
7/24	13.89"	18.36"	7"
1/3	14.42"	18.76"	8"
3/8	15.00"	19.21"	9"
5/12	15.62"	19.70"	10"
11/24	16.28"	20.22"	11"
1/2	16.97"	20.78"	12"
13/24	17.69"	21.38"	13"
7/12	18.44"	22.00"	14"
5/8	19.21"	22.65"	15"
2/3	20.00"	23.32"	16"
17/24	20.81"	24.02"	17"
3/4	21.63"	24.74"	18"
19/24	22.47"	25.48"	19"
5/6	23.32"	26.23"	20"
7/8	24.19"	27.00"	21"
11/12	25.06"	27.78"	22"
23/24	25.94"	28.58"	23"
1	26.83"	29.39"	24"
25/24	27.73"	30.22"	25"
13/12	28.64"	31.05"	26"
9/8	29.55"	31.89"	27"
7/6	30.46"	32.74"	28"
29/24	31.38"	33.60"	29"
5/4	32.31"	34.47"	30"

ALLOWANCES IN LAYING OUT RAFTERS
(From lengths computed by using data in previous table)

Pitch of Roof	Allowances for 1" Ridge Board		Allowances for Jack Rafters for Hip and Valley Rafters	
	Common and Valley Jack Rafters	Hip and Valley Rafters	Top of Hip Jack Rafters and Bottom of Valley Jack Rafters	Total Allowance on Cripple Jack Rafters
1/12	3/8"	1/2"	1 1/8"	2 1/4"
1/8	3/8"	1/2"	1 1/8"	2 1/4"
1/6	3/8"	1/2"	1 1/4"	2 3/8"
5/24	3/8"	1/2"	1 1/4"	2 1/2"
1/4	3/8"	5/8"	1 1/4"	2 1/2"
7/24	3/8"	5/8"	1 3/8"	2 3/4"
1/3	1/2"	5/8"	1 3/8"	2 3/4"
3/8	1/2"	5/8"	1 3/8"	2 7/8"
5/12	1/2"	5/8"	1 1/2"	2 7/8"
11/24	1/2"	5/8"	1 1/2"	3"
1/2	1/2"	5/8"	1 5/8"	3 1/4"
13/24	1/2"	5/8"	1 3/4"	3 3/8"
7/12	5/8"	3/4"	1 3/4"	3 1/2"
5/8	5/8"	3/4"	1 7/8"	3 5/8"
2/3	5/8"	3/4"	1 7/8"	3 3/4"
17/24	5/8"	3/4"	2"	3 7/8"
3/4	5/8"	3/4"	2 1/8"	4 1/8"
19/24	3/4"	3/4"	2 1/8"	4 1/4"
5/6	3/4"	7/8"	2 1/4"	4 3/8"
7/8	3/4"	7/8"	2 3/8"	4 5/8"
11/12	3/4"	7/8"	2 3/8"	4 5/8"
23/24	3/4"	7/8"	2 1/2"	4 7/8"
1	7/8"	7/8"	2 5/8"	5 1/8"
25/24	7/8"	1"	2 5/8"	5 1/4"
13/12	7/8"	1"	2 3/4"	5 3/8"
9/8	7/8"	1"	2 7/8"	5 3/8"
7/6	1"	1"	2 7/8"	5 3/4"
29/24	1"	1"	2 7/8"	5 7/8"
5/4	1"	1 1/8"	3"	6"

MAXIMUM SPANS FOR RAFTERS NO. 1 COMMON

American Standard Lumber Sizes		Dist. on Cntr.	Maximum Clear Span, Plate to Ridge					
			So. Pine and Douglas Fir		Western Hemlock		Spruce	
Nominal	Net		Unplstd.	Plstd.	Unplstd.	Plstd.	Unplstd.	Plstd.
Roof Load of 30 lb. per Sq. Ft. Uniformly Distributed for Slopes of 20° or More								
2"×4"	1⅝"×3⅝"	16"	7'-8"	6'-10"	7'-4"	6'-6"	7'-0"	6'-2"
		24"	6'-3"	6'-0"	6'-0"	5'-8"	5'-9"	5'-5"
2"×6"	1⅝"×5⅝"	16"	11'-9"	10'-6"	11'-3"	10'-1"	10'-9"	9'-7"
		24"	9'-8"	9'-3"	9'-3"	8'-10"	8'-10"	8'-5"
3"×6"	2⅝"×5⅝"	16"	14'-10"	12'-3"	14'-1"	11'-9"	13'-6"	11'-1"
		24"	12'-3"	10'-10"	11'-9"	10'-4"	11'-1"	9'-10"
2"×8"	1⅝"×7½"	16"	15'-7"	14'-0"	15'-0"	13'-4"	14'-3"	12'-9"
		24"	12'-10"	12'-3"	12'-4"	11'-9"	11'-9"	11'-2"
3"×8"	2⅝"×7½"	16"	19'-5"	16'-1"	18'-7"	15'-5"	17'-9"	14'-7"
		24"	16'-1"	14'-3"	15'-5"	13'-7"	14'-9"	12'-11"
2"×10"	1⅝"×9½"	16"	19'-7"	17'-6"	18'-9"	16'-10"	17'-11"	15'-11"
		24"	16'-3"	15'-6"	15'-6"	14'-10"	14'-10"	14'-0"
2"×12"	1⅝"×11½"	16"	23'-6"	21'-2"	22'-6"	20'-3"	21'-6"	19'-3"
		24"	19'-6"	18'-8"	17'-10"	17'-10"	17'-10"	17'-0"

Note: Deflection limited to 1/360 of the span. Dead load figured to include weight of rafters, roof sheathing, and 2.5 lb. for wood shingle or three-ply ready-made roofing. For heavier roof finishes use rafter next size larger. Data supplied by National Lumber Manufacturers Association.

STANDARD STOCK DOUGLAS FIR PLYWOOD SIZES—
OVERLAID PLYWOOD

Grade	Width In.	Length In.	Thickness (Inches)
A-High Density (Exterior)	36	96	5/16 (3-ply)
A-A High Density (Exterior)	48	96	⅜ (3-ply), ½ (5-ply), 9/16 (5-ply), ⅝ (5-ply), ¾ (5-ply), ⅞ (7-ply), 1 (7-ply)
B-B High Density (Exterior)	36, 48	96	Same as above
B-B High Density Ext. Conc. Form	48	96	½ (5-ply), 9/16 (5-ply), ⅝ (5-ply), ¾ (5-ply)
B-B Medium Density (Exterior)	36, 48	96	Same as for grade AA above
B-B Medium Density Ext. Conc. Form	48	96	½ (5-ply), 9/16 (5-ply), ⅝ (5-ply), ¾ (5-ply)

SPECIFICATIONS FOR ROUND REDWOOD PILING

All piles shall be cut from sound, live, old-growth trees and shall be free from injurious ring shakes, rot, loose or unsound knots, large knots, or other defects which will materially impair their strength or durability.

Piles shall have a uniform taper from butt to tip. The diameter 3' from the butt shall not be more than 1" less than the diameter at the butt. Piling shall be so straight that a straight line drawn from the center of the butt to the center of the tip will not deviate more than 1" for each 10' of length from the center of the pile; as, for example, a 50' pile may not deviate more than 5" from a straight line drawn from the center of the butt to the center of the tip.

Piles with short or reverse bends or kinks cannot be accepted. No piling with spiral grain which makes one complete turn in 40' or less will be accepted.

All piles, unless otherwise specified, shall be peeled free from bark. All knots and limbs shall be neatly trimmed flush with the surface of the pile. All butts and tips shall be cut square with clean end cuts prior to inspection. No piling will be accepted which has been cut for a period of time greater than 1 year prior to shipment. Untreated piles shall not be stored in salt water.

Piling shall not show less than an average of seven annual rings per linear inch measured radially over an area beginning 2" from the center of the heart of the stick.

DIMENSIONS

The diameter of a pile at any section shall be the average diameter of the stick at that section at right angles to the length, but no diameter of any cross section which is more than 10% larger than the least diameter shall be considered in obtaining an average. The maximum average diameter shall not be greater, and the minimum average diameter shall not be less, than the dimensions set forth in the following table for the various lengths.

CALIFORNIA REDWOOD

Length, Feet	Maximum Diameter Butt, In.	Minimum Diameter Butt, In.	Minimum Diameter Tip. In.
85 or more	22	16	7½
75 to 84	21	15½	8
65 to 74	20	15	8
55 to 64	20	14½	8
45 to 54	19	14	8
35 to 44	18	13½	8
25 to 34	18	13½	8
15 to 24	17	13½	8

The average diameter of heartwood in California redwood piles shall not be less than that shown in the following tables at the section being considered except that: When a heartwood diameter greater than the minimum required for any length is furnished the average thickness of sapwood may be increased by ⅓ of the increase in diameter of heartwood over that required. In no case shall the average thickness of sapwood be more than 2¼". When excess sapwood is permitted under above provision the contractor shall tap such excess of sap to the maximum thickness specified in the following table under all braces.

Average Outside of Section, In.	Average Outside Diameter of Heartwood, In.
20 to 22	16
18	15
16	13
14	11
12	9½
10	7½
8	5½

BEAMS, GIRDERS, AND STRINGERS 5" AND THICKER
(Modulus of elasticity for all uses, all grades, and all locations, 1,200,000)

Grade	Conditions Which Are		
	Always Dry	Wet and Dry	Usually Wet
	Lb. per Sq. In.		
Superstructural:			
Extreme fiber in bending	1,707	1,422	1,138
Horizontal shear	93	93	93
Compression across grain	267	160	133
Prime structural:			
Extreme fiber in bending	1,494	1,245	995
Horizontal shear	82	82	82
Compression across grain	267	160	133
Select structural:			
Extreme fiber in bending	1,322	1,100	880
Horizontal shear	70	70	70
Compression across grain	267	160	133
Heart structural:			
Extreme fiber in bending	1,150	960	768
Horizontal shear	56	56	56
Compression across grain	267	160	133

JOIST, PLANK, AND STRINGERS 4" AND THINNER
(Modulus of elasticity for all uses, all grades, and all locations, 1,200,000)

Grade	Conditions Which Are		
	Always Dry	Wet and Dry	Usually Wet
	Lb. per Sq. In.		
Superstructural:			
Extreme fiber in bending	2,133	1,420	1,135
Horizontal shear	93	93	93
Compression across grain	267	160	133
Prime structural:			
Extreme fiber in bending	1,707	1,182	945
Horizontal shear	82	82	82
Compression across grain	267	160	133
Select structural:			
Extreme fiber in bending	1,280	938	758
Horizontal shear	70	70	70
Compression across grain	267	160	133
Heart structural:			
Extreme fiber in bending	1,024	805	644
Horizontal shear	56	56	56
Compression across grain	267	160	133

RECOMMENDED GRADES OF CALIFORNIA REDWOOD

The use, grade, and size information given in the following tables may be helpful in selecting the proper quality, size, and item of California redwood lumber to use for given purposes. In each case the grade recommended is that which is suitable for the highest class of construction. In some cases an alternate grade is shown. For less expensive construction the next lower grade, as determined from the grade descriptions, may be used.

BUILDINGS, LIGHT-FRAMED OR LIGHT-JOISTED CONSTRUCTION, ROUGH CARPENTRY

Use-Item	Grades Recommended
Sleepers (foundation timbers), sills, beams, joists, rafters, headers	No. 1 heart dimension or timbers
Posts and columns	No. 2 dimension or timbers
Studs, plates, caps, bucks, headers, screeds	No. 2 dimension
Ribbon boards, collar beams, ridge boards, bracing, furring, grounds, bridging	No. 2 boards or dimension
Subflooring	No. 2 boards or No. 3 boards
Wall sheathing, roof sheathing, pitched	No. 2 boards
Roof decking, flat	No. 1 heart dimension or No. 2 dimension
Lath, wall and ceiling	No. 1 lath
Stair stringers or carriages	No. 1 heart dimension
Cellar and attic stair treads and risers	No. 1 heart boards
Shingles, roof	No. 1 grade, 5/2" or thicker

BUILDINGS, HEAVY-FRAMED, HEAVY-JOISTED, OR HEAVY-TIMBERED MILL CONSTRUCTION, ROUGH CARPENTRY

Use-Item	Grades Recommended
Sleepers (foundation timbers), posts and columns, sills	Select all-heart structural
Beams, girders, purlins, joists, headers	Dense, select all-heart structural
Studs, plates, caps, bucks, partitions, plank or laminated	No. 1 heart dimension
Ribbon boards, collar beams, ridge boards, bracing, furring, grounds	No. 2 boards or dimension
Subflooring	No. 1 heart boards or No. 2 boards
Screeds	No. 2 dimension
Factory flooring, plank or laminated	No. 1 heart dimension
Wall sheathing, roof sheathing, pitched	No. 2 boards
Roof decking, flat	No. 1 heart dimension
Lath, wall and ceiling	No. 1 lath
Stair stringers or carriages	No. 1 heart dimension
Cellar and attic stair treads and risers	No. 1 heart boards or dimension
Shelving, warehouse	No. 1 heart dimension
Shingles, roof	No. 1 Grade 5/2" or thicker
Roof trusses, compression and tension members	Dense, select all-heart structural

BUILDINGS, FRAMED, JOISTED, OR HEAVY-TIMBERED MILL CONSTRUCTION, EXTERIOR FINISHED CARPENTRY AND MILLWORK

Use-Item	Grades or Qualities Recommended
Siding:	
Beveled, drop, square-edge, vertical	Clear heart or A siding or finish
Battens	Clear heart or A and better moldings
Log-cabin siding	Clear, or No. 1 heart dimension
Shingles, side wall	No. 1 grade shingles
Trim:	
Cornice, trim, corner boards, half timbering	Clear heart finish
Crown and bed moldings, drip cap, water table	Clear heart or A and better moldings
Store fronts, painted	Clear heart finish or A finish
Store fronts, metal-covered	No. 1 heart boards
Gutters	Clear heart gutters
Window and door:	
Frames	Clear heart or A frames
Sash, windows and storm doors, standard	Clear heart
Doors, residential, garage, or warehouse	Clear heart finish or A finish
Shutters, blinds, screens	Clear heart
Porch:	
Ceiling	A ceiling or B ceiling
Flooring	Edge-grain clear heart flooring
Stair stringers or carriages	No. 1 heart boards or No. 1 heart dimension
Stair treads and risers	Clear heart finish
Columns, built-up and solid	Clear heart or A columns
Newel posts, railings, balustrades	Clear heart or A and better moldings

BUILDINGS, FIREPROOFED CONSTRUCTION, ROUGH AND EXTERIOR CARPENTRY

Use-Item	Grades or Qualities Recommended
Rough carpentry:	
Furring, grounds, subflooring	No. 1 heart boards
Screeds, roof decking, flat	No. 1 heart dimension
Lath, wall and ceiling	No. 1 lath
Exterior Finished Carpentry:	
Half timbering-frames, window and door-sash, windows and storm doors, shutters, blinds, screens	Clear heart

BUILDINGS, FRAMED, JOISTED, HEAVY-TIMBERED MILL, OR FIREPROOFED CONSTRUCTION, INTERIOR FINISHED CARPENTRY, MILLWORK AND CABINETWORK

Use-Item	Grades or Qualities Recommended	
	Natural Finishes	Painted Finishes, or Items Not Exposed to View
Finish and trim (stock):		
Door and window	Clear heart finish, or A and better moldings	A finish, or A and better moldings
Moldings, casing and base	A and better moldings	A and better moldings
Ceiling, partition	Clear heart	A or B
Closet lining	A finish, or A ceiling	B finish or B ceiling
Ceiling beams, solid and built-up	Clear heart finish	A finish

BUILDINGS, FRAMED, JOISTED, HEAVY-TIMBERED MILL, OR FIREPROOFED CONSTRUCTION, INTERIOR FINISHED CARPENTRY, MILLWORK AND CABINETWORK (*Continued*)

Use-Item	Grades or Qualities Recommended	
	Natural Finishes	Painted Finishes, or Items Not Exposed to View
Doors	Clear heart doors	A-finish doors
Shelving	A finish	No. 1 heart boards
Counter tops	Clear heart finish	Clear heart finish
Special millwork and cabinetwork:		
Built-in equipment (service)	A-finish	B finish or N . 1 heart boards
Built-in equipment (ornamental)	Clear heart finish	A finish
Paneling, stiles, rails	Clear heart finish	A finish
Finished sheathing for interior walls	Clear heart finish	A finish
Stairs (stock patterns):		
Stringers or carriages (exposed to view), treads, risers, skirting	....	Clear heart finish or A finish
Railings, newel posts, and balustrades	....	A and better moldings

ROUGH CARPENTRY-NAIL CHART

Size Timbers	Timber Items	Wire nails, Common, Per 1,000 F.b.m., Lb.		
		10d	20d	30d
6 × 8	Girder stay, bracing and Supports	4	6	
2 × 6	Floor joist	8.5	13	
2 × 8	Floor joist	6.5	9.5	
2 × 10	Floor joist	6.5	10	
2 × 12	Floor joist	6	9	
3 × 8	Floor joist	6	....	12
3 × 10	Floor joist	4.5	....	10
2 × 4	Studs and plates (side wall)	9.5	14.5	
2 × 4	Ceiling beams	6	8.5	
2 × 4	Partition studs, plates, and shoes	11.5	5	
4 × 6	Corner posts	3	6	
2 × 4	Common and hip rafters	15	22	
2 × 6	Common and hip rafters	11	16.5	
2 × 8	Common and hip rafters	8	16.5	
2 × 10	Common and hip rafters	6.5	14.5	
2 × 4	Collar beam	....	24	
2 × 6	Collar beams	....	26	
2 × 8	Collar beams	....	22	
2 × 6	Porch frame and floor joist	3	26	
2 × 8	Porch frame and floor joist	3	26	
2 × 4	Porch ceiling beams	5	11.5	
2 × 6	Porch ceiling beams	7.5	16.5	
2 × 8	Double porch plates	3	26	

ROUGH CARPENTRY-NAIL CHART (*Continued*)

Size Timbers	Timber Items	Wire nails, Common, Per 1,000 F.b.m., Lb.		
		10d	20d	30d
2 × 10	Double porch plates	3	26	
2 × 4	Porch rafters	13	32	
2 × 6	Porch rafters	15	32	
2 × 4	Fire blocks	60	....	
2 × 6	Fire blocks	60	....	
2 × 4	Partition bridging	73.5	....	
1½ × 3	Floor bridging	124	....	

ESTIMATING FLOORING

Wood flooring is usually 25/32" thick and is tongued and grooved, or "matched." The most commonly used face widths are 2¼" and 3¼", which are made from 3" and 4" stock and are figured on that basis.

Since there are a number of grades of flooring on the market, care should be taken by the estimator to determine correctly the price for the kind of work specified. Oak flooring, for example, is classified as white or red and comes in different grades such as clear, select, No. 1 or No. 2 common. Similar grading is applied to maple or pine flooring. In laying wood floors, there is a considerable loss of surface due to stock requirements and cutting on the job, so that it becomes necessary to add between one-third and one-half to the net floor area in order to obtain the net number of feet. board measure.

In roughly estimating flooring, find the net area of the floor, and then add 20% for 1" × 6" underflooring laid diagonally, and about 25% per 1,000 f.b.m.

A more accurate method is to add the allowances given in the following table:

PERCENTAGES TO BE ADDED TO WOOD FLOORS
(Determine area of floors and add percentages shown.)

Nominal Size, In.	Actual Size, In.	Per Cent Added for Waste	Number of Ft. Flooring for 100-Sq.-Ft. Floor (Per Cent for Waste Added)
1 × 1	3/8 × 7/8	16⅔	117
1 × 2	3/8 × 1½	33⅓	133
1 × 2¼	13/16 × 1½	50⅓	150
1 × 2½	3/8 × 2	25	125
1 × 2¾	13/16 × 2	37½	137½
1 × 3	13/16 × 2¼	33⅓	133
1 × 4	13/16 × 3¼	25	125

AMOUNT OF SURFACE 1,000 FT. OF FLOORING WILL COVER AND QUANTITY OF NAILS REQUIRED TO LAY IT

Nominal Size, In.	Actual Size, In.	Will Cover Sq. Ft. Floor	Nail Spacing, In.	Nails Required, Lb.	Kind of Nails
1 × 2	3/8 × 1½	750	8	20	4d casing
1 × 2¼	13/16 × 1½	667	12	70	8d coated casing
1 × 2½	3/8 × 2	800	8	17	4d casing
1 × 2¾	13/16 × 2	727	12	56	8d coated casing
1 × 3	13/16 × 2¼	750	10	64	8d coated casing
1 × 4	13/16 × 3¼	800	10	29	8d coated casing

NAILS REQUIRED FOR SQUARE-EDGED AND MATCHED BOARDS

Items	Dimensions, In.	Wire Nails (8d) per 1,000 F.b.m., Lb.
Sheathing laid straight	⅞ × 8*	31
Sheathing laid straight	⅞ × 6*	40
Sheathing laid straight	⅞ × 6†	31
Sheathing laid diagonal	⅞ × 8*	30
Sheathing laid diagonal	⅞ × 6*	39
Rough floor laid straight	⅞ × 6†	30
Rough floor laid straight	⅞ × 8*	31
Rough floor laid straight	⅞ × 6*	40
Rough floor laid diagonal	⅞ × 6†	31
Rough floor laid diagonal	⅞ × 8*	30
Rough floor laid diagonal	⅞ × 6*	39
Roof boards, gable roof	⅞ × 6†	30
Roof boards, hip roof	⅞ × 6*	31
Roof boards, gable roof	⅞ × 6*	31
Roof boards, hip roof	⅞ × 6†	23
Roof lath, gable roof	⅞ × 6†	23
Roof lath, hip roof	⅞ × 2†	49
Plancier	⅞ × 3†	32
Plancier	⅞ × 4*	30

*Tongue and groove.
†Straight edge.

PLYWOOD THICKNESSES, SPANS AND NAILING RECOMMENDATIONS
(Plywood Continuous Over 2 or More Spans; Grain of Face Plys Across Supports)

Application	Recommended Thickness	Max. Spacing of Supports (C. to C.)	Nail Size and Type	Nail Spacing	
				Panel Edges	Inter-mediate
Subflooring	½" (a)	16" (b)	6d Common (c)	6"	10"
	⅝" (a)	20"	8d Common (c)	6"	10"
	¾" (a)	24"	8d Common (c)	6"	10"
	2.4.1	48"	8d Ring Shank (c)	6"	6"
Under-lay ment	⅜" (d)		6d Ring Shank or Cement Coated	6"	8" Each Way
	⅝"		8d Flathead		

(a) Provide blocking at panel edges for carpet, tile, linoleum or other non-structural flooring. No blocking required for 25/32" strip flooring.
(b) If strip flooring is perpendicular to supports ½" can be used on 24" span.
(c) If resilient flooring is to be applied without underlayment, set nails ¹⁄₁₆".
(d) FHA accepts ¼" plywood.

If supports are not well seasoned, use ring-shank nails.

Courtesy of Douglas Fir Plywood Association, Tacoma 2, Washington.

DATA ON SANDING WOOD FLOORS PER 100 SQ. FT.

Type of Floor, Where Used	Type of Workmanship	Number of Sq. Ft. Finish and Edge per 8-hr. Day
In homes, apartments, offices, etc.	Ordinary	800 to 900
Storerooms, auditoriums, other, large floor areas	Ordinary	1,000 to 1,250, new floor
Residences, apartments, hotels, stores, office buildings	First-class work - about 4 cuts necessary	400 to 500

WOOD SHINGLES

To determine the number of shingles required for a roof or wall, find the number of square feet of roof or wall surface, then divide by 100 to get the number of squares. The number of shingles in each square depends on the shingle exposure (see table). It is customary to add 10% to this amount for waste in cutting.

Wood shingles are sold 4 bundles to the square. Since 4 bundles of shingles make up 1 square, it is possible to order 3 bundles which will cover ¾ of a square, 2 bundles which will cover ½ of a square, or one bundle which will cover ¼ of a square, When the number of squares are known, the number of bundles can easily be determined.

EXAMPLE: A roof contains 12¼ squares; how many bundles are required?

SOLUTION: Twelve squares contain

$12 \times 4 = 48$ bundles
¼ square contains 1 bundle
Total number of bundles $= 48 + 1 = 49$ bundles

NUMBER OF SHINGLES AND NAILS
REQUIRED PER SQUARE

Length, In.	Distance Laid to Weather	Number of Shingles per Square	Weight of Nails (3d) per Square	Weight of Nails (4d) per Square
14	4	900*	3.75 lbs.	6.50 lbs.
15	4½	800*	3.25 lbs.	6.00 lbs.
16	5	720*	3.00 lbs.	5.25 lbs.
18	5½	655*	2.75 lbs.	4.75 lbs.
20	6	600*	2.50 lbs.	4.25 lbs.
22	6½	554*	2.25 lbs.	3.75 lbs.
24	7	515*	2.00 lbs.	3.25 lbs.

*Add 10% to above amounts for waste.

AREA IN SQUARE FEET COVERED BY ONE BUNDLE OF WOOD SHINGLES*

Width of Exposure	Length of Shingles		
	16"	18"	24"
3½	17½		
4	20	17½	
4½	22½	20	
5	25	22½	
5½	27½	25	
6	30	27	20
6½	32½	29	22½
7	35	31	24
7½	37½	34	25
8	40	36	26½
8½	42½	38	28
9	45	40	30
9½	47½	43	31½
10	50	45	33
10½	52½	47	35
11	55	50	36½
11½	57½	52	38
12	60	54	40
12½	....	57	41½
13	....	59	43
13½	....	61	45
14	....	63	46½
14½			48
15			50
15½			51½
16			53

***Courtesy of the Red Cedar Shingle Bureau.**

SIDE-WALL COVERING CAPACITIES (PER FOUR-BUNDLE SQUARE) IN SQUARE FEET FOR THE VARIOUS-SIZED SHINGLES

Single Course				Double Course			
Exposure, In.	16" Shingle	18" Shingle	24" Shingle	Exposure Double Course	16" Shingle	18" Shingle	24" Shingle
7	140						
7½	150	......	......	11½	226		
8	......	146	......	12	238	218	
8½	......	154	......	13	......	236	
9	......	......	120	14	......	254	
10	......	......	132	15	......	......	200
11	......	......	146	16	......	......	212
11½	......	......	152				

Note: Quantities shown under "Double Course" are for each course.

RUNNING-INCH METHOD OF ESTIMATING WOOD SHINGLES

Another method of figuring the quantity of shingles required is the running-inch method.

The length of each row in inches times the number of rows, as determined by the exposure and size of area, will give the running inches of shingles required for roofs or side walls. Allowances should be made for starters, hips, ridges, and valleys, and deductions made for window and door openings.

"RUNNING INCHES" PER FOUR-BUNDLE SQUARE OF SHINGLES

Length, In.	Thickness— Green	Number of Running Inches per 4-bundle square	
		Green	Dry
16	5 butts, 2	2,960	2,880
18	5 butts, 2¼	2,664	2,620
24	4 butts, 2	1,996	1,920

TIME REQUIRED TO LAY ONE SQUARE (100 SQ. FT.) WOOD SHINGLES
(Based on using 2 nails to each shingle and 10% waste)

Class of work	Mechanic	Number Laid per 8-hr. Day	Distance Shingles Laid to Weather					
			4"	4¼	4½	5	5½	6
Plain gable or hip roofs	Carpenter	2,000 to 2,200	3.8	3.6	3.4	3.0	2.8	2.5
	Shingler	2,750 to 3,000	2.8	2.6	2.5	2.3	2.0	1.9
Difficult gable roofs, cut up with gables, dormers, hips, valleys	Carpenter	1,700 to 1,900	4.5	4.2	4.0	3.6	3.3	3.0
	Shingler	2,200 to 2,500	3.4	3.2	3.0	2.7	2.5	2.3
Difficult hip roofs, steep English roofs, hips, valleys	Carpenter	1,300 to 1,500	5.7	5.4	5.1	4.6	4.2	3.8
	Shingler	2,000 to 2,200	3.8	3.6	3.4	3.0	2.8	2.5
Shingles laid irregularly or with staggered butts on plain roofs	Carpenter	1,700 to 1,900	4.5	4.2	4.0	3.6	3.3	3.0
	Shingler	2,400 to 2,700	3.1	3.0	2.8	2.5	2.3	2.1
Shingles laid irregularly or with staggered butts on roofs of difficult construction	Carpenter	1,100 to 1,300	6.7	6.3	6.0	5.4	4.9	4.5
	Chingler	1,600 to 1,800	4.7	4.5	4.2	3.8	3.4	3.1
Shingles with Thatched Butts*	Shingler	800 to 1,000	8.9	8.4	8.0	7.1	6.5	6.0
Plain side walls	Carpenter	1,300 to 1,500	5.7	5.4	5.1	4.6	4.2	3.8
	Shingler	1,700 to 1,900	4.5	4.3	4.0	3.6	3.3	3.0
Difficult side walls, having bays windows, breaks	Carpenter	1,100 to 1,250	6.8	6.4	6.0	5.4	5.0	4.5
	Shingler	1,400 to 1,650	5.2	5.0	4.6	4.2	3.8	3.5

*Shingles with thatched butts require 25% more shingles than when laid regularly.

HOW TO DETERMINE ROOF AREAS

Roof areas can quickly be found by multiplying the horizontal distance between the lower ends of the gable by the length of the roof. This will yield the area of the horizontal flat. By multiplying the flat area by the percentage factor in the following table for the various roof pitches the surface area of the roof is found.

For example: In the following illustration, multiply distance *AB* by distance *BC*; then multiply by the percentage factor in table to get the area of the entire roof surface.

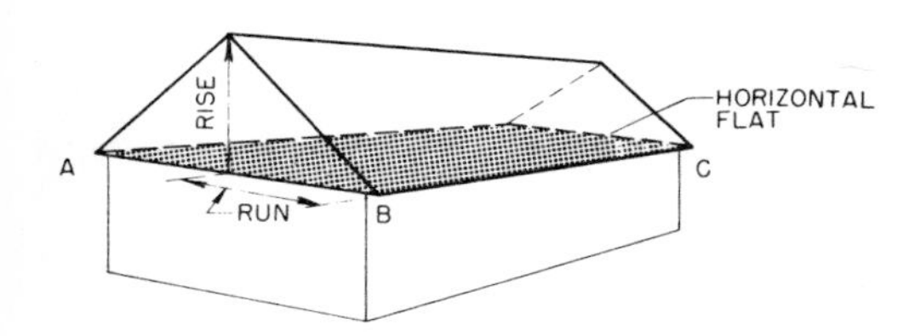

FINDING ROOF AREA

PERCENTAGE FACTOR FOR ROOF AREAS

Rise, In.	Run, In.	Pitch	Percentage Factor
2	12	1/12	1.014
3	12	1/8	1.031
4	12	1/6	1.054
5	12	5/24	1.083
6	12	1/4	1.118
7	12	7/24	1.158
8	12	1/3	1.202
9	12	3/8	1.250
10	12	5/12	1.302
11	12	11/24	1.357
12	12	1/2	1.414
13	12	13/24	1.474
14	12	7/12	1.537
15	12	5/8	1.601
16	12	2/3	1.667
17	12	17/24	1.734
18	12	3/4	1.803
19	12	19/24	1.873
20	12	5/6	1.943
21	12	7/8	2.016
22	12	11/12	2.088
23	12	23/24	2.162
24	12	1	2.236
25	12	25/24	2.311
26	12	13/12	2.387
27	12	9/8	2.463
28	12	7/6	2.538
29	12	29/24	2.615
30	12	5/4	2.693

GLOSSARY OF LUMBER ABBREVIATIONS

B&Btr., B and better
BCB2S, Bead and center bead two sides
Ch., Choice
Clr., Clear
Col., Colonial
DB1S, Double-beaded one side
DB2S, Double-beaded two sides
DV1S, Double V one side
DV2S, Double V two sides
D.F., Douglas fir
D&M, Dressed and matched (Tongue and groove)
EM, End-matched
FG, Flat grain
Hdw. Pat., Hardwood pattern
IWP, Idaho white pine
MG, Mixed grain
O.C. (o.c.), On center
PP or Pond. Pine, Ponderosa pine
Qual., Quality
Sel., Select
Sel. Merch., Select merchantable
Shlp., Shiplap
Spcl. Pat., Special pattern

GLOSSARY OF LUMBER ABBREVIATIONS (*Continued*)

Std., Standard
Jointed, Square edge (S4S)
Str., Sterling
Sup., Supreme
S4S, Surfaced four sides
S3SN1E, Surfaced three sides nosed one side
S2S, Surfaced two sides
S2S & CM, Surfaced two sides and center-matched
T&G, Tongue and groove
Utl., Utility
VCV2S, V and center V two sides
VG, Vertical grain
W C Hem., West Coast hemlock
W R Cedar, Western red cedar

ASPHALT SHINGLES

Most asphalt shingles are packed in two, three, or four bundles to the square. One bundle of a 2-bundle square will cover 50 sq. ft., one bundle of a 3-bundle square will cover 33⅓ sq. ft., and one bundle of a 4-bundle square will cover 25 sq. ft.

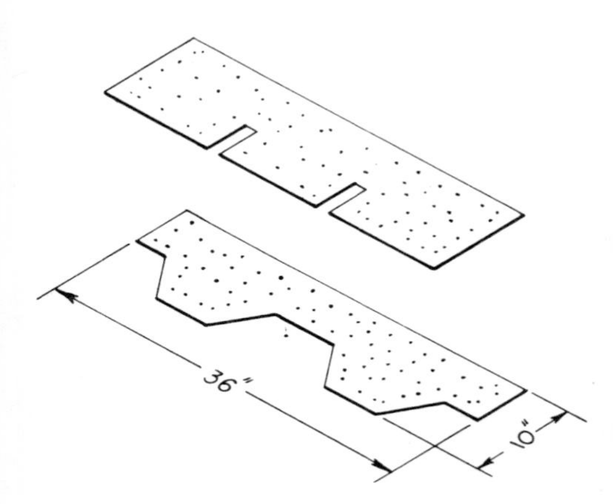

ASPHALT STRIP SHINGLES
80 SHINGLES PER SQUARE
1½ LB. OF 1½" NAILS

After the area of the roof is determined, add 5% of this area to the roof area for waste, and then divide the total area by 100 to get the number of squares: Once the number of squares is known, the number of bundles of shingles can easily be determined from the above. Count any fractional part of a bundle as one bundle.

DATA ON ASPHALT SHINGLES*

Product	Weight per Square, Lb.	Head Lap, In.	Exposure, In.	Shingles per Square	Bundles per Square
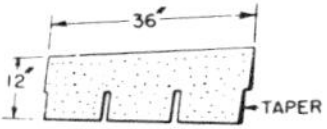 TAPERED STRIP SHINGLES	275	2	5	80	3
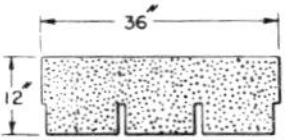 GIANT OVERLAY STRIP SHINGLES	250	2	5	80	3
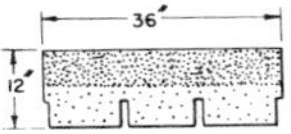 12" THIKBUT SHINGLES	210	2	5	80	3
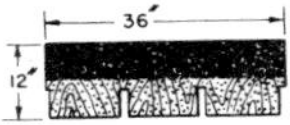 12" CEDARTEX THIKBUT SHINGLES	210	2	5	80	3
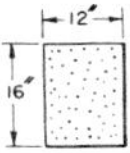 SUPER GIANT INDIVIDUAL SHINGLES	325	6	5	226	4
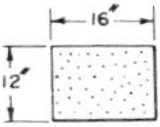 DUTCH LAP GIANT INDIVIDUAL	162	2	10 × 13	113	2
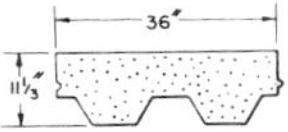 11⅓" HEXAGON SHINGLES	167	2	4⅔	86	2

*Packed two bundles per square in the South.

To find the extra number of shingles required for the starter course, get the combined total length of the starter courses of a roof, and divide by the length of a single course which can be laid by one bundle of shingles. The following table will give the number of feet covered by one bundle of shingles.

LENGTH IN FEET OF A SINGLE COURSE OF SHINGLES WHICH CAN BE LAID BY ONE BUNDLE OF ROOFING SHINGLES

Width of Exposure, In.	2-bundle squares	3-bundle Squares	4-bundle Squares	5-bundle Squares
4	150	100	75	60
4½	133	88	66	53
5	120	80	60	48
5½	109	72	54	43
6	100	66	50	40
6½	92	63	46	37
7	85	57	42	34
7½	80	53	40	32
8	75	50	37	30

To find the number of additional bundles to allow for waste on valleys and hips, divide the combined length of these valleys and hips by:

30 for 2-bundle square
20 for 3-bundle square
15 for 4-bundle square

Each square of shingles requires about 1½ lb. of 1½"-long galvanized nails. It requires about 3 hr. labor time to lay one square of shingles.

ROOFING FELT

Before shingles are laid, a heavy roofing felt is required. This felt is made of the same material as the shingle. It is furnished in rolls of 500 sq. ft. Figure about 2 hr. labor time to lay one square of roofing felt.

AVERAGE SIZES AND WEIGHTS OF ROLL ROOFING

Material	Width, In.	Area per Roll, Sq. Ft.	Weight per Roll, Lbs.
15 lb. asphalt or tarred felt	32 and 36	432	60
30 lb. asphalt or tarred felt	32 and 36	216	60
Stater's felt	36	500	32
Sheathing felt	36	500	35
Red rosin-sized sheathing paper	36	500	20, 25, 30, 40
Roll roofing: tack both sides	36	108	45 to 65
Mineral-surfaced	36	108	55 to 90

ASBESTOS SHINGLES

Asbestos shingles are sold by the square, covering 100 sq. ft. of surface. These shingles are rigid and are made of asbestos fiber and portland cement. They are furnished in numerous styles, sizes, and colors. In estimating the quantities required for a roof, determine the total square feet of roof area, and divide by 100 to get the number of squares required.

Lengths of eaves must be measured to obtain the number of linear feet of starters required. Also, the length of hips, valleys, and ridges must be found as either shingles or special ridge roll will be required.

For shed or gable roofs which do not contain hips or valleys, add 5% to the area to allow for waste. Add 2% additional to the 5% if the roof has hips or valleys.

DATA ON ASBESTOS-CEMENT SHINGLES

Product	Weight per Square, Lb.	Head Lap, In.	Side Lap, In.	Exposure, In.	Shingles per Square
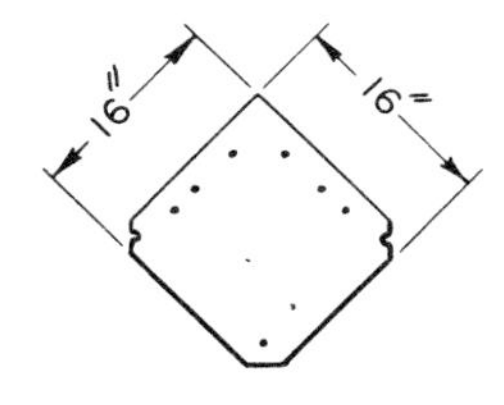 HEXAGONAL SMOOTH FINISH	265	.....	3	13 × 13	86
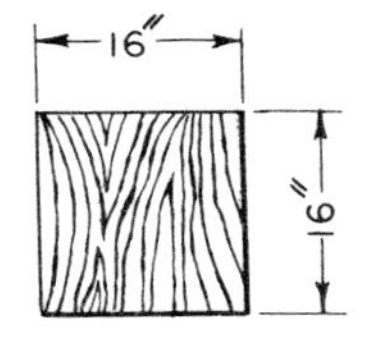 DUTCH LAP WOODGRAIN FINISH	280	3	4	12 × 13	92

DATA ON ASBESTOS-CEMENT SIDING

Product	Weight per Square, In.	Head Lap, In.	Side Lap, In.	Exposure, In.	Shingles per Square
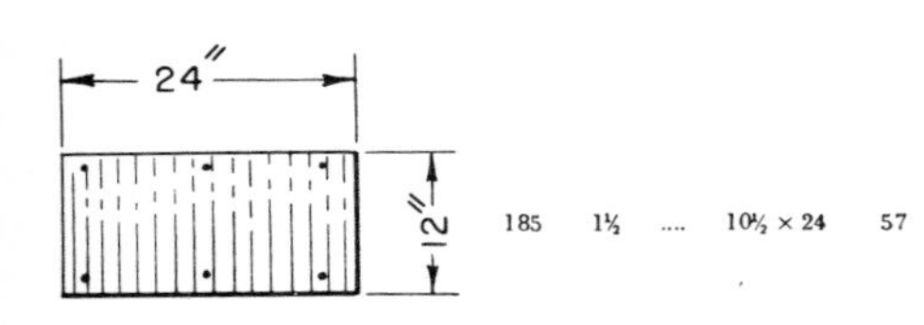 TAPERTEX STRAIGHT EDGE	185	1½	....	10½ × 24	57
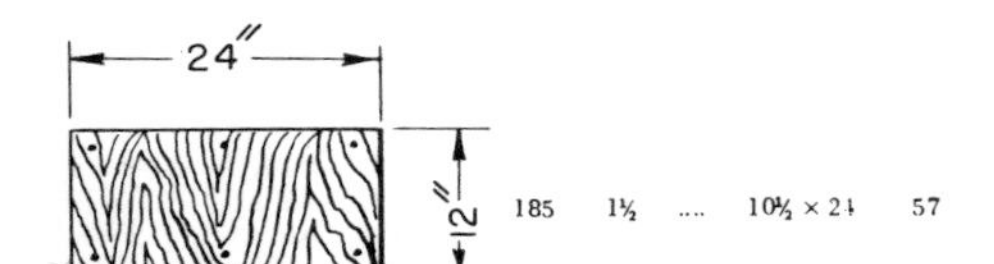 WAVELINE WOODGRAIN FINISH	185	1½	....	10½ × 24	57
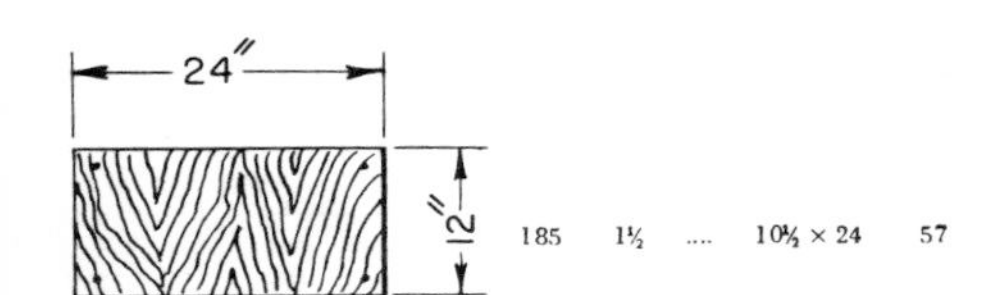 STRAIGHT EDGE WOODGRAIN FINISH	185	1½	....	10½ × 24	57

CORRUGATED TRANSITE DIMENSIONS AND WEIGHTS

Corrugated transite sheets have corrugations with a 4.2" pitch and a depth of 1½". The thickness is approximately 7/16" at the ridge and valley of corrugations and approximately 5/16" on tangent, an average thickness of 3/8". Sheets are furnished 42", or 10 corrugations, wide. Standard lengths are listed below. The number of square feet covered by these lengths is also included.

Length	Sq. Ft. Area	Length	Sq. Ft. Area
0'-6"	1.75	6'-0"	21.00
1'-0"	3.50	6'-6"	22.75
1'-6"	5.25	7'-0"	24.50
2'-0"	7.00	7'-6"	26.25
2'-6"	8.75	8'-0"	28.00
3'-0"	10.50	8'-6"	29.75
3'-6"	12.25	9'-0"	31.50
4'-0"	14.00	9'-6"	33.25
4'-6"	15.75	10'-0"	35.00
5'-0"	17.50	10'-6"	36.75
5'-6"	19.25	11'-0"	38.50

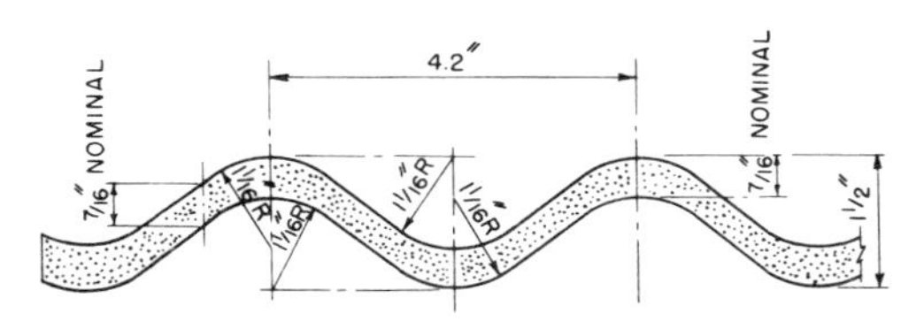

CORRUGATED TRANSITE SHEET DIMENSIONS

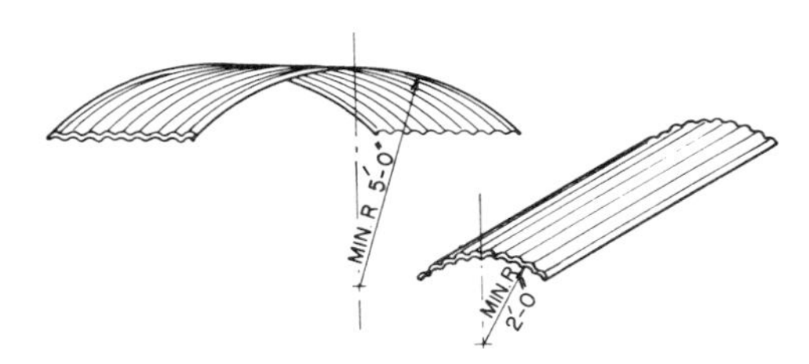

CURVED CORRUGATED SHEETS

POUNDS OF NAILS REQUIRED PER SQUARE FOR MATERIALS USED IN ROOFING AND SIDING

Kind of Material	Kind of Nails	Lb. of Nails Required
Wood shingles, 16"	3 d shingle	4 lb. per 4-bundle square
Wood shingles, 16"	4 d shingle	6 lb. per 4-bundle square
Wood shingles, 16"	5 d shingle	7 lb. per 4-bundle square
Asphalt strip shingles	1" roofing	2½ lb. per square
Asphalt strip shingles	1½" roofing	3 lb. per square
Roll roofing	1" roofing	1¼ lb. per roll
Roll roofing	1½" roofing	1½ lb. per roll
Roll siding	1" roofing	¼ lb. per square
	1¼" black siding	2 lb. per square
Roll siding	1½" roofing	¼ lb. per square
	1¾" black siding	2½ lb. per square
Insulation siding	1¾" black siding	2 lb. per square
Asphalt strip siding	1½" siding	1 lb. per square
	1½" zinc	1 lb. per square
Asbestos roofing shingles	1½" needle point galvanized	2½ lb. per square
Asbestos roofing shingles	2" needle point galvanized	3 lb. per square
Asbestos siding shingles	1¼" needle point galvanized	¾ lb. per square
Asbestos siding shingles	2" needle point galvanized	1 lb. per square

ESTIMATING QUANTITIES OF INSULATION

Insulation may be ordered by giving the number of square feet needed or by giving the number of packaged units in which the insulation is delivered. Following are the steps necessary for figuring insulating quantities:

1. Find the distance around the building, using exterior dimensions.
2. Multiply the distance around the building by the ceiling height for a one-story building. For a two-story building figure the height from the first floor to the second-floor ceiling.
3. If the combined area of all the exterior openings is more than 5% of the total area, find the combined area of these openings, omitting any which have less than 3 sq. ft. If deductions for openings are to be made, subtract the combined area of these openings from the total area.
4. Add 8% to the total area to allow for waste. If the ceiling is to be insulated, find its area, and add 8% for waste.
5. If one kind of insulation is to be used on both side walls and ceiling, find the sum of the quantities of insulation and divide this sum by the number of square feet of insulation which can be covered by one packaged unit, counting any fractional part of a unit as one additional.

TYPES OF INSULATION

The principal types of insulation available are as follows:

1. The blanket type.
2. The batt type.
3. The granulated or fluffed type.
4. The rigid type.

The *blanket type* is a kind of fluffed material placed between two layers of waterproof paper. It is furnished in rolls or in folded layers and in widths to fit between studs and joists of various spacings.

The *batt type* consists of a thick pad of fluffed material attached to a waterproof backing. It is manufactured in sizes to fit between the studs and joists of various spacings. This insulation is delivered in cartons. The average density of wood batts is 4¾ to 5 lb. per cu. ft.

Granulated insulation consists of loose material which can be poured or blown between the studs or joists. It is packed in bags. A bag contains about 4 cu. ft. with a covering capacity of 17 sq. ft. for 3" thick spread, and 14.6 sq. ft. for 3½" thick spread.

The *rigid type* not only serves as insulating material but is also used as a plaster base or in place of wood sheathing as a structural member. This rigid insulation is manufactured in 18" widths and 32" or 48" lengths. It is packed in cartons for delivery.

SIZES AND COVERING CAPACITY OF MINERAL- OR ROCK-WOOL BATTS

Size of Batts, In.	Thickness, In.	Sq. Ft. per Batt		Number of Batts per 100 Sq. Ft.	
		Including 2 × 4's	Excluding 2 × 4's	Including 2 × 4's	Excluding 2 × 4's
15 × 24	2	2.67	2.50	38	40
15 × 24	3	2.67	2.50	38	40
15 × 48	2	5.33	5.00	19	20
15 × 48	3	5.33	5.00	19	20
19 × 24	2	3.33	3.17	30	32
19 × 24	3	3.33	3.17	30	32
19 × 48	2	6.67	6.33	15	16
19 × 48	3	6.67	6.33	15	16
23 × 24	2	4.00	3.84	25	26
23 × 24	3	4.00	3.84	25	26
23 × 48	2	8.00	7.67	13	13
23 × 48	3	8.00	7.67	13	13

MINERAL- OR ROCK-WOOL INSULATION

Rock-wool insulation is made of rock or silica melted at high temperatures and blown into fine, threadlike particles, which when cooled form a wool-like substance that is used for insulation between studs, ceiling joists, and roof rafters and also for deadening between floors and partitions. The average density of rock-wool batts is 4¾ to 5 lb. per cu. ft.

AMOUNTS OF WOOL INSULATION REQUIRED PER SQUARE FOOT

Solid Studs or Joists		Lb. per Sq. Ft. of Superficial Area
Size	Spacing	
2" × 3"	16" centers	.88
	24" centers	.93
2" × 4"	16" centers	1.22
	24" centers	1.27
2" × 6"	12" centers	1.85
	16" centers	1.89
	24" centers	1.97
2" × 8"	12" centers	2.43
	16" centers	2.51
	24" centers	2.63
2" × 10"	12" centers	3.09
	16" centers	3.19
	24" centers	3.33
2" × 12"	12" centers	3.73
	16" centers	3.86
	24" centers	4.03

INSULATING VALUES OF VARIOUS MATERIALS

Material	D	K	R	Authority
Palco wool (Redwood bark fiber)	5.0	.255	3.92	J. C. Peebles
Redwood bark fiber*	5.0	.284	3.52	U.S. Bureau of Standards
Corkboard	10.6	.300	3.33	U.S. Bureau of Standards
Corkboard (low density)	7.0	.270	3.70	U.S. Bureau of Standards
Cork (regranulated)	8.1	.311	3.22	U.S. Bureau of Standards
Rock cork	15.6	.328	3.05	U.S. Bureau of Standards
Celotex	13.5	.330	3.03	J. C. Peebles
Insulite	15.9	.330	3.03	J. C. Peebles
Thermax	26.4	.458	2.18	J. C. Peebles
Glass wool	1.5	.266	3.76	J. C. Peebles
Mineral wool	10.5	.310	3.22	U.S. Bureau of Standards
Rock wool	14.28	.280	3.57	U.S. Bureau of Standards
Sawdust	12.8	.410	2.44	U.S. Bureau of Standards
Shavings (planer)	8.8	.410	2.44	U.S. Bureau of Standards
Sheep's wool	8.5	.338	2.96	Musselt
Balsam wool	2.2	.270	3.70	U.S. Bureau of Standards
Hair felt	17.0	.264	3.78	U.S. Bureau of Standards
Kapok (dry zero)	.98	.240	4.17	U.S. Bureau of Standards
Redwood lumber	25.5	.570	1.75	G. F. Gebhardt (Armour Institute of Technology)
Balsa wood	8.8	.380	2.63	U.S. Bureau of Standards
Cypress	29.0	.670	1.49	U.S. Bureau of Standards
Oak	38.0	1.020	.98	U.S. Bureau of Standards
White pine	32.0	.780	1.28	U.S. Bureau of Standards
Brick (low density)	120.0	5.000	.200	Average
Brick (high density)	....	9.200	.109	Average
Concrete	145.0	8.000	.125	Average
Concrete (cinder)	97.35	4.86	.206	Average
Concrete (blocks, typical 8")†	....	1.00	1.00	Average
Hollow tile (typical 4")†	....	1.00	1.00	Average
Plaster (cement)	....	8.00	.125	University of Illinois
Wood lath and plaster†	....	2.50	.400	F. B. Rowley
Asphalt-composition roofing†	....	6.50	.154	J. C. Peebles
Air film—ordinary f_0 15 miles per hr. wind†	....	6.00	.166	Accepted average for ordinary calculation
Air film—ordinary f_1 still air†	....	1.65	.605	
Air spaces, ¾" and over†	....	1.10	.91	F. B. Rowley
Aluminum foils, 4 curtains, five ¾" spaces (3¾")†	....	.09	11.66	F. B. Rowley
Surface of ground (for floors)†‡	....	2.000	.50	Accepted practice
1" sheathing, building paper, and lap siding†	....	.50	2.00	F. B. Rowley

D = density, lb. per cu. ft.

K = conductivity, B.t.u. per hr. per in. of thickness per sq. ft. per deg. F. temperature difference on either side of the material except where marked (†). A British thermal unit (B.t.u.) is the amount of heat required to raise the temperature of one pound of water one degree Fahrenheit.

R = resistance, or the number of degrees Fahrenheit difference in temperature required on opposite sides of a material for 1 B.t.u. to pass through one square foot of one inch thickness in one hour, which is the reciprocal of the conductivity K of a material ($R = 1/K$).

*Test made on Redwood bark fiber. (See Bulletin No. 243—Nov. 1930.)

†Indicates values "C" for actual thickness—not per inch of thickness.

‡Ground temperatures usually equal mean annual temperatures in any given locality.

FOAMGLAS INSULATION FOR FLOORS

Foamglas is composed of inert air hermetically sealed in glass. It is light and strong and impervious to deteriorating elements. The following illustrations show how foamglas should be applied to insulate standard floor construction in many types of industrial buildings.

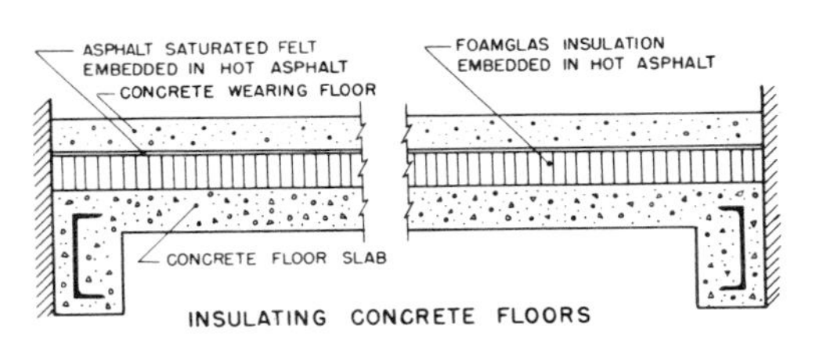

INSULATING CONCRETE FLOORS

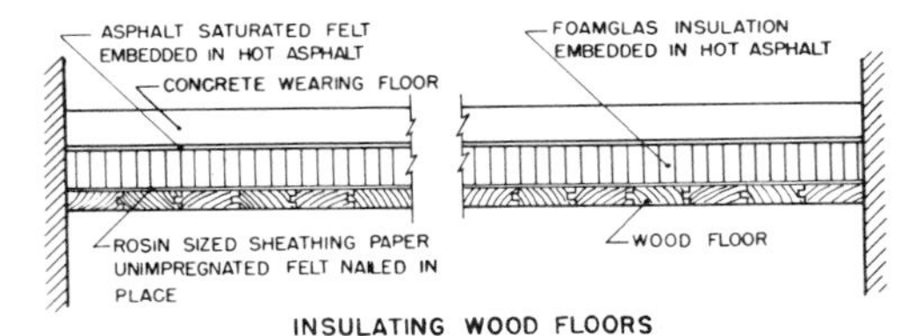

INSULATING WOOD FLOORS

SIZES AND PACKING OF FOAMGLAS

Standard Sizes	Pieces per Carton	Sq. Ft. per Carton	Approximate Weight per Carton, Lb.
12 × 18 × 2	12	18	40.5
12 × 18 × 3	8	12	35.0
12 × 18 × 4	6	9	38.5
12 × 18 × 5	6	9	43.5

PREPARATION OF FLOORS FOR FOAMGLAS

WOOD FLOORS: The surface of wood floors to be insulated shall be reasonably smooth and level, without depressions. All loose and springy boards should be properly nailed in place. Floor should further be broomed clean, free from dirt and loose material, and thoroughly dry before proceeding. Over all wood floors to be insulated apply a layer of rosin-sized sheathing paper or unsaturated felt, lapping the edges at least 3" and nailing along the edges to hold in place until the insulation is laid.

CONCRETE FLOORS: All new concrete floor slabs shall be thoroughly cured, reasonably smooth and level, and free of all loose particles and dirt.

Foamglas is laid in paralleled courses, staggered to break joints. All joints should be tightly butted.

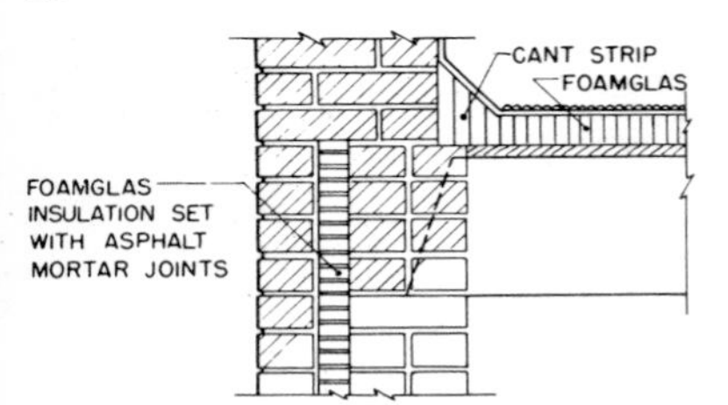

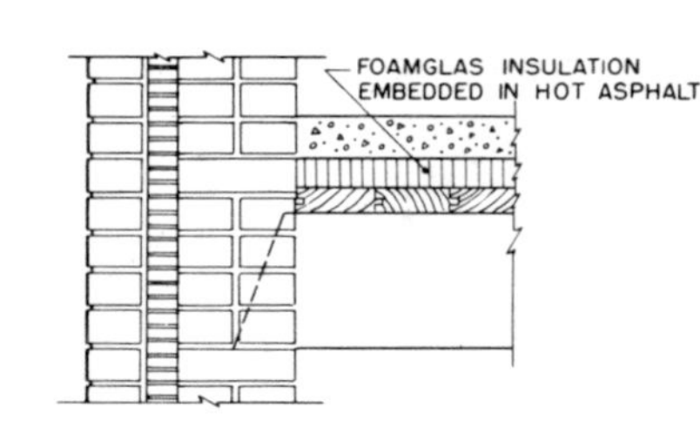

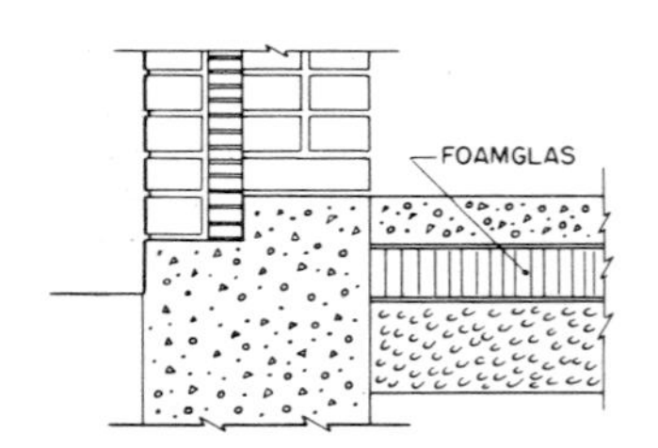

FOAMGLAS INSULATION IN LOAD BEARING TYPE CORE WALL CONSTRUCTION

OVER WOOD AND CONCRETE FLOORS: Foamglas over wood and concrete floors should be firmly embedded in hot asphalt and shall be laid progressively as the hot asphalt is mopped on the floor. Avoid mopping large areas that cannot be covered with insulation before the hot asphalt cools.

After foamglas is laid, mop on a layer of 15-lb. asphalt saturated felt with hot asphalt. When mopping the insulation and before the felt is laid, the hot asphalt shall be slushed into and fill all joints.

STRUCTURAL COMPOSITION MATERIAL

COMPOSITION SHEATHING: Composition sheathing may be applied in place of regular wood sheathing. It is fabricated from the tough fibers of hardy northern woods, felted together into rigid boards containing innumerable minute air spaces. Panels are $\frac{25}{32}$" thick and 4' wide by 8', $8\frac{1}{2}$', 9', 10', and 12' long. These panels are applied vertically or paralleled to the framing members. For horizontal sheathing, or at right angles to the framing members, 2' × 8' panels with V joints or shiplap joints are used. Some of these sheets are coated with high-melting-point asphalt, which forms a moisture-resistant surface and retards moisture penetration.

Composition sheathing is usually applied to wood studs, under wood siding, shingles, stucco, or brick veneer. Because of the large sized sheets used in the asphalt treatment, building paper is not ordinarily also employed.

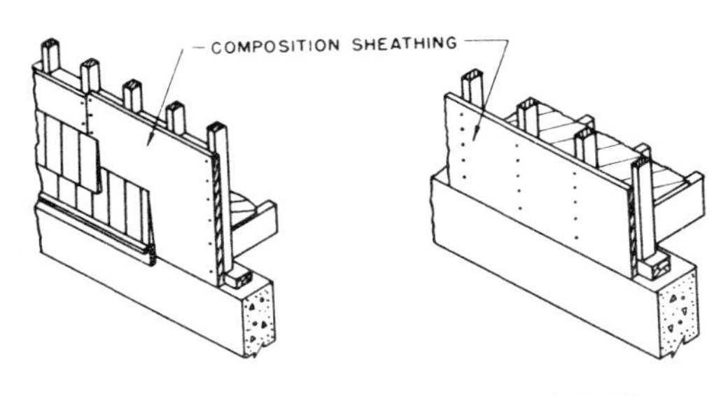

APPLICATION OF COMPOSITION SHEATHING

PLYWOOD SIDING

Plywood siding panels are made in 4 x 8 ft. panels. Thicknesses come in $\frac{3}{8}$ to $\frac{5}{8}$ inch. The $\frac{3}{8}$ in. is a 3-ply panel while the $\frac{5}{8}$ in. is a 5-ply panel. Panels are put together forming $\frac{1}{8}$ or $\frac{3}{8}$ in. wide channel grooves. A panel may have three vertical grooves 5/32 or $\frac{3}{8}$ in. wide by 3/32 or $\frac{1}{4}$ in. deep. This gives the panel the appearance of three individual boards. Some designs are of the board and batten treatment. Narrow vertical strips of wood are nailed onto the flat panel. The strips are called battens. The batten strips give the appearance of an individual or single board.

ALUMINUM SIDING

Aluminum siding comes in vertical and horizontal design. The vertical siding may be 12 in. wide. It is made to give an appearance of two individual 6 inch vertical panels. The horizontal siding is made with a weather exposure of 4, 5, and 8 inches. The 8 in. exposure is more common. Sidings come unbacked, polystyrene backed, and fiberboard backed for insulation. Siding length is 12'-6". Surface finishes come in many designs and patterns. Some have a wood grain pattern in colors of white, tan, sandstone and fern green.

INSULATING TILE

Insulating tile is a fiber composition used for wall and ceiling fini
Most insulating tile is furnished with beveled edges and in sizes
12" × 12" squares to 16" × 32". Some of the trade names of this ma
are Insulite, Celatex, Temlock, and Nu-wood.

The tile may be applied direct to a wood backing or to furring stri
the walls and ceilings. When furring strips are used, the strips mu
spaced to accommodate the size tile used.

Nails should be driven through the tile at an angle of about 60°.
through the edge of the tile, but not through the bevel. Bury nailhead
low the surface, using nail set.

QUANTITIES AND LABOR COSTS OF APPLYING FIBER TILE TO WALLS AND CEILINGS WHERE TILE ARE NAILED IN PLACE*

Size of Tile, In.	Number Sq. Ft. per Tile	Number Tile per 100 Sq. Ft.	Number Placed per 8 Hr.	Number Sq. Ft. per 8 Hr.	Carpent Hr. pe 100 Sq.
12 × 12	1.0	100	180 to 200	180 to 200	4.3
12 × 24	2.0	50	95 to 105	190 to 210	4.0
16 × 16	1.78	56¼	120 to 130	215 to 235	3.7
16 × 32	3.56	28¼	85 to 90	295 to 315	2.8

*Add for wood furring strips or other necessary backing.

STEEL JOIST HANGERS*

Joist Size, In.	Bar Size, In.	Strength, Lb.	Weight, Each, Lb.	Pri Ea
2 × 6	1½ × ⅛	4,400	1-lb.	$.:
2 × 8	2 × ⅛	6,000	1¼-lbs.	.4
2 × 10	2½ × ⅛	7,500	2½-lbs.	.5
2 × 12	3 × ⅛	9,000	2¾-lbs.	.7
4 × 8	2 × ⅛	6,000	1⅞-lbs.	.4
4 × 10	2½ × ⅛	7,500	2¾-lbs.	.6
4 × 12	3 × ⅛	9,000	3¾-lbs.	.8

*Holes punched for nails.

ITEMIZED LIST FOR ESTIMATING LUMBER FOR A FRAME, BRICK-VENEER, BRICK, OR STONE DWELLING OR APARTMENT HOUSE

Sills
End sills
Side sills
Cross sills
Trimmers
Basement posts

Joists
First-floor joists
Second-floor joists
Third-floor joists
Ceiling joists
Roof joists

Studding
Wall studding
Partition studding
Gable studding
Plate and braces
Ribbons

Flooring
Subfloors
Attic floors
Flooring, 1st and 2nd stories
Flooring, all other stories
Deafening felt or paper
Mineral wool

Roof Timbers
Rafters (common)
Hip rafters
Valley rafters
Jack rafters
Purlins
Trusses and braces
Collar beams

Enclosing Building
Shiplap or sheathing
Roof sheathing
Shingles
Dimension shingles

Siding
Paper
Bridging
Furring
Grounds
Lining sliding-door pockets
sash, doors
Rear steps

Cellar
Cellar stairs
Cellar doors
Cellar door frames
Cellar windows
Cellar window frames
Bulkhead door and frame

Cornice and Finish
Fascia, crown moulding
Plancier, finish or ceiling
Lookouts
Frieze, bed molding
Brackets
Ridgeboards
Corner boards

Lath
Metal lath
Rock lath

Window Frames
Side and head jambs
Back lining
Inside casing of frame
Sills and subsills
Outside casing
Water table and molding
Blind stops
Parting stops, brick mold
Lintels

Windows and Sash
Windows, 1st story
Casement sash
Windows, 2nd story
Windows, 3d story
Windows, other stories
Gable windows
Storm windows
Screen windows
Transoms

Doors
Front door
Front-door side lights
Side doors, French doors
Rear doors

Outside Door Frames
Side and head jambs
Transom bar
Doorsills
Outside casing
Water table and molding
Brick mold
Lintels

Inside Doors
Vestibule doors
Sliding doors
Astragals, banding doors
Attic doors
Storm doors
Screen doors
Closet doors

Interior Finish
Jamb lining
Window casing, mullion casing
Door casing
Head casing
Moldings, for head casing
Band molding
Window stool
Window apron and molding
Window stops
Window seats
Doorjambs
Sliding-door jambs
Transom bar
Doorstops
Plinth blocks
Head blocks
Base molding

ITEMIZED LIST FOR ESTIMATING LUMBER FOR A FRAME, BRICK-VENEER, BRICK, OR STONE DWELLING OR APARTMENT HOUSE
(*Continued*)

Interior Finish (*Continued*)
Base shoe or quarter round
Mantel trimming
Wainscoating and cap
Thresholds

Kitchen and Pantry Finish
Hook strips
Cupboards
Shelving
Drawers
Cupboard doors
Cupboard finish
Worktable
Shelf and braces

Closet Finish
Casing
Base and moldings
Base blocks
Shelving

Closet Finish (Continued)
Drawers
Hook strips

Stairs
Stair horses
Treads and risers
Molding
Platforms
Flooring
Wall strings
Main newels
Platform newels
Landing newels
Rails and shoe
Fillet
Easers
Balusters
Panelwork
Hall seat

PLASTERING AND LATHING

All plasterwork is measured by the square yard for flat surfaces and by the linear foot for moldings, corner beads, and corners. Many estimators make deductions for openings, others deduct only half of the openings; and still others do not consider any openings less than 21 sq. ft. The most common method and, at the same time, the easiest is not to make any deductions for openings at all but to allow a flat 7% of the gross area.

For example: In determining the number of square yards of plastering in a room 16'-0" × 20'-0" having 8'-0" ceiling height, take the perimeter of the room, such as 16 + 16 + 20 + 20 = 72 lin. ft. wall.

72 × 8 = 576 sq. ft. wall area
16 × 20 = 320 sq. ft. ceiling area

896 sq. ft. area wall and ceiling
896 ÷ 9 sq. ft. = 99.5 sq. yd.
7% of 99.5 = .07 × 99.5 = 6.96
Therefore, 99.5 − 6.96 = 92.54 net sq. yd.

A distinction must be made between lime, gypsum, and cement plaster. While gypsum plaster is used on gypsum blocks and plasterboards, cement plaster is used in places where fire resistance or water resistance is desired.

The first plaster coat, called the scratch coat, consists of lime, sand, and hair and is forced into the lath so that some of the material comes through and forms a key. Before it hardens, it is scratched with a tool or a piece of metal lath so that a rough surface is created which will act as a base for the next coat.

The second coat is the thickest of the three coats. It consists of lime and sand only and is troweled to a true surface with a wooden trowel called a float. The finest coat consists of lime and plaster of paris in a thickness of about ⅛", and it is troweled with a steel trowel to a hard, smooth finish, ready for painting or wallpaper covering.

QUANTITIES OF MATERIALS REQUIRED FOR PLASTERING PER 100 SQ. FT. OF WALL SURFACE

Plaster Coats	Materials	Applied on Plaster Bases		Applied on Masonry	
		Composition Lath or Insulation Plaster	Metal Lath	Clay Tile Masonry	Concrete Masonry
Brown and scratch coats	Gypsum plaster, 100-lb. sacks	1.11	2.00	1.33	
	Sand, 1:2 mixture, cu. yd.	.09	.17	.11	
Brown and scratch coats	Masonry cement, 100-lb. sacks	1.17	2.05	1.41	1.41
	Sand, 1:2 mixture, cu. yd.	.10	.17	.12	.12
Lime putty finish	Hydrate of lime, 50-lb. sacks	.67	.67	.67	.67
	Gaging plaster, 100-lb. sacks	.17	.17	.17	.17
Sand-float finish	Prepared sand-float finish, 100-lb. sacks	.89	.89	.89	.89
	Gypsum plaster, 100-lb. sacks	.75	.75	.75	.75
	Sand, 1:2 mixture, Cu. yd.	.06	.06	.06	.06
Trowel finish	Prepared trowel finish, 100-lb. sacks	.89	.89	.89	.89

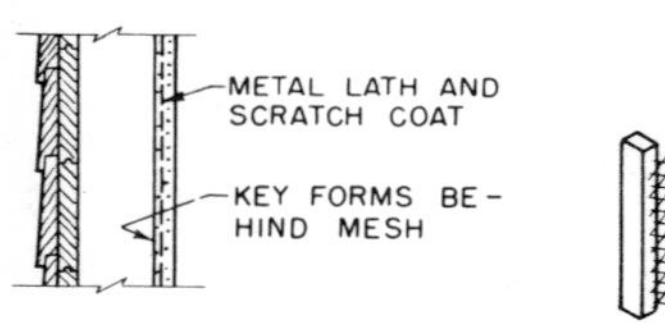

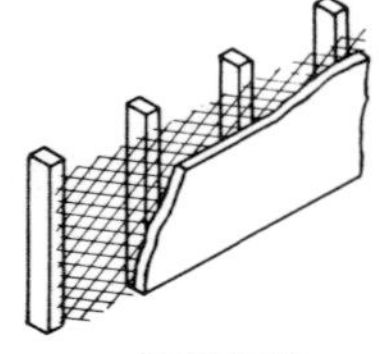

SECTION

PICTORIAL

WIRE LATH AND PLASTER

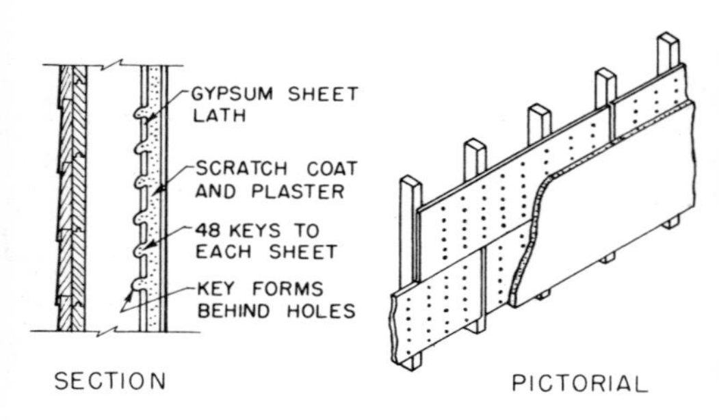

SHEET LATH AND PLASTER

NUMBER OF 100-LB. SACKS OF GYPSUM CEMENT PLASTER REQUIRED PER 100 SQ. YD. OVER THE FOLLOWING PLASTERING SURFACES

Metal Lath, ⅝" Grounds	Gypsum Lath, ¾" Grounds	Brick or Clay Tile Walls, ½" to ⅝" Grounds	Pyrobar Gypsum Tile, ½" Grounds
16 to 22, sacks, sanded 1:2, 1:3	9 to 11 sacks, sanded 1:2, 1:3	10 to 12 sacks, sanded 1:3	8.5 to 10 sacks, sanded 1:3
For Each ⅛" in Thickness, Add or Subtract			
3 sacks	2.75 sacks	2.5 sacks	2.5 sacks

NUMBER OF 100-LB. SACKS OF PREPARED OR SANDED PLASTER REQUIRED PER 100 SQ. YD. OVER THE FOLLOWING PLASTERING SURFACES

Metal Lath, 5/8" Grounds	Gypsum Lath, 3/4" Grounds	Brick or Clay Tile Walls, 1/2" to 5/8" Grounds	Gypsum Tile, 1/2" Grounds
45 to 57 sacks	25 to 27 sacks	36 to 45 sacks	28 to 31 sacks
For Each 1/8" in Thickness, Add or Subtract			
12 sacks	6.25 sacks	8 sacks	7.5 sacks

NUMBER OF 100-LB. SACKS OF WOOD-FIBER PLASTER REQUIRED PER 100 SQ. YD. OVER THE FOLLOWING PLASTERING SURFACES

Metal Lath, 5/8" Grounds	Gypsum Lath, 3/4" Grounds	Brick or Clay Tile Walls, 1/2" to 5/8" Grounds	Gypsum Tile, 1/2" Grounds
31 to 36 sacks, unsanded	17 to 18 sacks, unsanded	12 to 14 sacks	10 to 11 sacks
		Equal parts of sand to be added	
For Each 1/8" in Thickness, Add or Subtract			
8 sacks	4.5 sacks	3 sacks	2.5 sacks

MATERIALS OF PLASTER AND PLASTER PRODUCTS

Plaster consists of a cementing material combined with an aggregate, water and fibers. The aggregates are generally sand, perlite, vermiculite, and wood fibers. There are four types of cementing materials used to make plaster—Gypsum; Keen's cement; lime; cement. Gypsum plaster is not only used as a plastering material, but, also to make gypsum board and gypsum block.

Gypsum is found in certain kinds of rock. These are heated to 340 to burn off the moisture in them. The rocks are then ground to a fine powder that is called Plaster of Paris. Keen's cement is derived from gypsum. When gypsum is heated to 1,400, all its water content is driven off. The result is hard-burned gypsum. By grinding and pulverizing it a dense water resistant plaster is obtained. It is called Keen's cement.

Lime is made by crushing limestone. Limestone is a rock consisting mainly of magnesium and calcium carbonate. By burning the stone in kilns to 2,000, carbon dioxide is driven off and only dry solids remain. These are pulverized into fine powder called quicklime. When mixed with water large quantities of heat are given off. This is known as the slaking process. Lime that is slaked is called hydrated lime. All lime is slaked before it can be used.

COVERING CAPACITY OF LIME PLASTER

Lime plaster is proportioned by volume, 1 part of well-aged lime putty to 2 parts of dry sand, and the necessary hair or fiber for the scratch coat.

For the brown coat use 1 part of lime putty to 3 parts of dry sand and hair or fiber as needed.

COVERING CAPACITY OF 1 CU. YD. OF LIME PLASTER WITH THE FOLLOWING PLASTERING SURFACES

Type of Coat	Metal Lath, 5/8" Grounds	Brick or Tile, 1/2" Grounds
Scratch coat	50 to 60 sq. yd.	
Brown coat	70 to 80 sq. yd.	60 to 75 sq. yd.

ESTIMATING THE AMOUNT OF METAL LATH

A number of different kinds of metal lath are manufactured, the most common of which are the flat rib metal lath, the diamond metal lath, and the metal rib lath.

Diamond metal lath is manufactured in sheets 24 or 27" wide and 96" long. The 24" width is packed in 9-sheet bundles having a total area of 16 sq. yd. The 27" width is packed in 10-sheet bundles containing 20 sq. yd. of material.

In estimating the amount of metal lath required for a room, multiply the total linear feet of the wall by the height of the wall to get the number of square feet. If the room contains windows and doors, do not deduct for such openings. This extra amount will replace the waste in cutting. For ceilings, multiply length by width, and do not make any deductions.

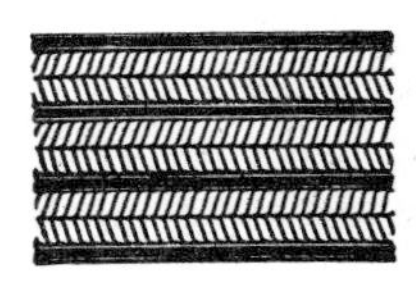

FLAT RIB METAL LATH
COURTESY OF NATIONAL GYPSUM CO.
BUFFALO, N. Y.

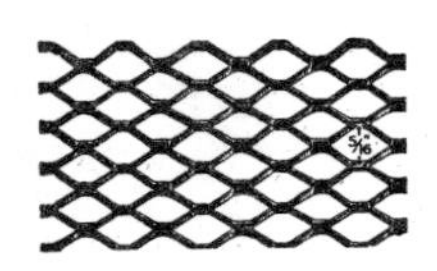

DIAMOND METAL LATH   METAL RIB LATH
COURTESY OF NATIONAL GYPSUM CO.
BUFFALO, N. Y.

If 24" × 96" sheets are used, divide the total square feet required for the room by 16, because there are 16 sq. ft. in one 24" × 96" sheet.

If 27" × 96" sheets are used, divide the total square feet required for the room by 18, because there are 18 sq. ft. in one 27" × 96" sheet.

To find the number of bundles, divide the area of the room by 9 to find the number of square yards. Divide the number of square yards by 16 for 9-sheet bundles of the 24" × 96" size or by 20 for 10-sheet bundles of the 27" × 96" size.

CORNER BEADS

Corner beads are used around openings and on corners of plastered walls to prevent chipping. Following is an illustration of typical corner beads. The stock lengths of corner beads are 8', 9', 10', and 12'.

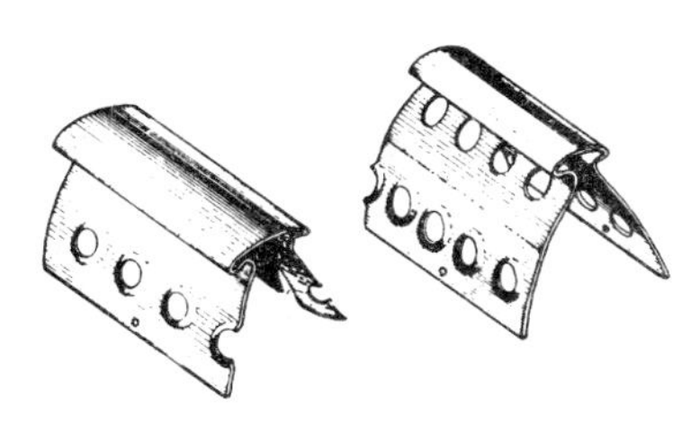

TYPICAL CORNER BEADS

To estimate corner beads, find the combined length of outside corners in all rooms to be plastered. If any two rooms are connected by a plastered archway, multiply the distance round the archway by 2, and add this number by linear feet of corner beads to any previously found.

If corner bead is to be ordered in stock lengths, select any suitable stock length, and divide the combined length of bead required by the stock length desired, counting any fractional part of a piece as one additional length.

HOW TO ESTIMATE GYPSUM LATH

To estimate gypsum lath, find the total area of the surfaces of walls and ceilings and deduct all window and door openings. Divide the area of one panel into the net area of the surfaces to be covered. The result is the number of panels of the size selected.

POUNDS OF NAILS REQUIRED FOR METAL LATH AND COMPOSITION PLASTER BASES

Kind of Lath	Kind of Nails	Lb. of Nails Required
Composition plaster base	1¼" plasterboard	7½ lb. per 1,000 sq. ft.
Insulation plaster base	1¼" plasterboard	7½ lb. per 1,000 sq. ft.
Steel lath 24" × 96"	1¼" roofing	2 lb. per 9-sheet bundle
	1½" roofing	2 lb. per 9-sheet bundle
Steel lath 27" × 96"	1¼" roofing	2½ lb. per 10-sheet bundle
	1½" roofing	2½ lb. per 10-sheet bundle

MAXIMUM ALLOWABLE SPACING OF SUPPORTS IN INCHES FOR VARIOUS TYPES OF METAL AND WIRE LATH

Type of Lath	Minimum Weight of Lath, Lb. per Sq. Yd.	Vertical Supports			Horizontal Supports	
		Wood	Metal Solid Partition	Other	Wood or Concrete	Metal
Flat expanded	2.5	16	16	12	0	0
Metal lath	3.4	16	16	16	16	13½
Flat rib	2.75	16	16	16	16	12
Metal lath	3.4	19	24	19	19	19
⅜" rib	3.4	24	....	24	24	19
Metal lath	4.0	24	....	24	24	24
Sheet metal lath	4.5	24	24	24	24	24
Wire lath	2.48	16	16	16	13½	13½
V-stiffened Wire lath	3.3	24	24	24	19	19
Wire fabric	......	16	0	16	16	16

APPROXIMATE COSTS OF METAL LATH

Type of Lath	Finish	Weight, Lb. per Sq. Yd.
Expanded diamond mesh (flat)	Painted	2.5
Expanded diamond mesh (flat)	Painted	3.4

APPROXIMATE COSTS OF RIB METAL LATH

Type of Lath	Finish	Weight, Lb. per Sq. Yd.
Herringbone or diamond mesh	Painted	3.4
Herringbone or diamond mesh	Painted	4.0
Rib metal lath	....	2.75
Rib metal lath	....	3.4

GYPSUM SHEATHING

Gypsum sheathing is ½" thick, 2 ft. wide and 6 ft. 8 in., 8 and 10 ft. long to fit supports 16 in. on centers. Gypsum sheathing has V-joint edges on the long dimension to provide a tight fit at the unsupported joints. It is applied with its length at right angles to the studs. It is made of a solid sheet of gypsum encased in a strong, fibrous, water-resisting covering. The sides and ends are treated to resist moisture. The nails used are 1¾ in. long, No. 10½ gauge, galvanized flat head roofing nails, spaced 4 in. on centers, except under wood siding and stucco, 8 inches. About 14 to 21 lbs. of nails are required per 1000 sq. ft.

GYPSUM WALL BOARD

This material is available in regular wall board, as an insulating wallboard, as a firestop wallboard, as a backer board, and as a gypsum sound deadening board. Thicknesses range from ¼ to ⅝ in. in increments of ⅛ inches. Widths are 4 ft. Lengths range from 6 to 12 ft.

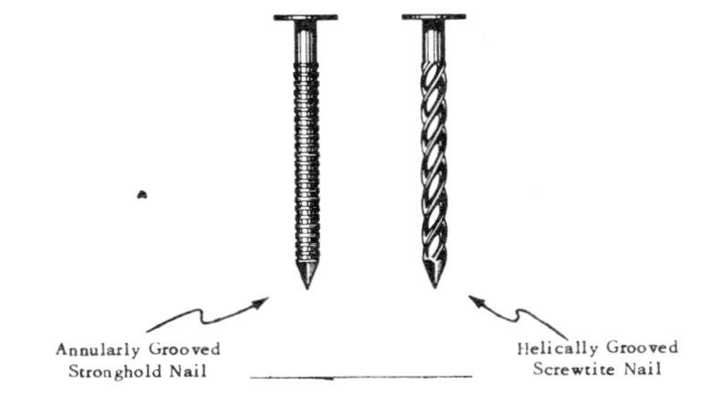

DATA ON THREADED NAILS

Two types of nails, the helically grooved and the annularly grooved, have recently been mass-produced so that they can compete in price with the plain-shank nail (see illustration 58a).

The helically grooved nail has a long-pitch shank that permits it to screw its way into the wood. The nail displaces the fibers and forms a thread in the wood. This compresses the surrounding fibers and increases the frictional resistance between the wood and the shank, and thus the holding power of the nail.

The annularly grooved nail has numerous grooves encircling the shank and when driven into the wood forces the fibers into the annular groovelike wedges, to be released only when the wood is destroyed.

Careful and extensive research has shown that helically grooved and annularly grooved nails, although harder to drive into wood, offer a greater load capacity and are harder to withdraw.

STRENGTH FACTORS OF THREADED NAILS

Nail Property	Nail Metal	Nail Length, In.	Plain-shank Nail, Lb. Resistance	Helically Grooved Screwtite Nail, Lb. Resistance	Annularly Grooved Stronghold Nail, Lb. Resistance
Withdrawal Resistance	Bright Steel	$\frac{7}{8}$	108	133	157
		1	97	159	164
	Galvanized steel	$\frac{7}{8}$	127	146	161
		1	112	143	169
	Aluminum alloy	$\frac{7}{8}$	100	127	133
		1	102	149	164
Lateral load-carrying capacity	Bright steel	$\frac{7}{8}$	232	316	246
		1	220	267	259
	Galvanized steel	$\frac{7}{8}$	224	248	117
		1	217	248	228
	Aluminum alloy	$\frac{7}{8}$	199	258	239
		1	209	269	230

NAILS FOR VARIOUS TYPE OF WORK

Type of Work	Nail Size	Type of Work	Nail Size
Framing	20d	Finishing	10d
	16d		8d
	10d		6d
	8d		4d
	6d	Siding	7d
Casing	10d	Shingles	3d
	6d	Asphalt Roofing	$1\frac{1}{8}$"-13ga
Flooring	8d	Lath	1"-15ga

The "Penny" system of designating nails originated in England. Two explanations are offered by the Southern Pine Association as to how this curious designation came about. One is that the sixpenny, fourpenny, tenpenny etc., nails derived their names from the fact that one hundred cost sixpence, fourpence, etc. The other explanation is that one thousand tenpenny nails, weighed ten pounds. The ancient as well as the modern abbreviation for penny is "d", being first letter of the Roman coin Denarius. The same abbreviation in early history was used for the English pound in weight. The "penny" at any rate has persisted in the nail industry.

ADVANTAGES OF THREADED NAILS

Threaded nails are somewhat more expensive than the regular shank nails. The advantages gained by the use of threaded nails will justify this increased cost. The following can be achieved by the use of threaded nails.

1. Framing can offer four to six times greater lateral thrust resistance.
2. Flooring and stair treads can be laid more easily and quicker.
3. Siding may be fastened in such a way that the nails will not "creep" or "pop". Thus the siding remains firmly fastened to the structure.
4. Assurance may be had that none of the wood or asbestos shingles come loose, since the use of grooved nails results in a 250% greater holding power.
5. Assurance may be had in fastening asphalt roofing shingles that the nails retain the shingles continuously tightly fastened as a result of a 50% greater axial withdrawal resistance of the grooved nails in comparison with the plain shank nails.
6. Through threaded nails the plaster lath may be permanently firmly fastened to the frame work because of the 40% increased holding power of the nail.
7. When used for casing and finishing nails, the wood work should never come loose.

ESTIMATING WOOD FURRING STRIPS

Wood furring is estimated by the linear foot. A carpenter should place 500 to 550 lin. ft. of furring per 8-hr. day, or 62.5 to 68.75 lin. ft. per hr.

Where wedging and blocking out are required to produce a level surface for plaster, the carpenter should place 160 to 200 lin. ft. of strips per 8-hr. day, or 20 to 25 lin. ft. per hr.

PLACING WOOD GROUNDS

A carpenter should place 540 to 590 lin. ft. of wood grounds per 8-hr. day, or 67.5 to 73.75 lin. ft. per hr., provided the grounds are nailed tight to wood furring strips, door or window openings, without wedging or blocking to make them absolutely straight.

Where wood grounds must be absolutely straight, plumb and level to the plaster, for first-class workmanship, a carpenter will place 160 to 200 lin. ft. per 8-hr. day.

WOOD FURRING STRIPS ON MASONRY WALLS

Before interior wall finishes are applied to a masonry wall, it is generally the practice to fur the wall. Furring strips are usually 1" × 2", made of wood, and secured to the wall with special fasteners. Furring strips are spaced 12" to 16" apart depending on the type of lath which is to be applied.

When wire lath or rock lath is applied to the furring strips, the 1" space between the lath and the wall may be used for insulation material.

The term "contact furring" implies that the furring strips are fastened to the wall. Sometimes, however, an entire wall may be supported by furring strips not in contact with the wall to provide space for plumbing, heating, or electrical conduits. In this case furring strips may be as large as 2" × 3" or 2" × 4" studs.

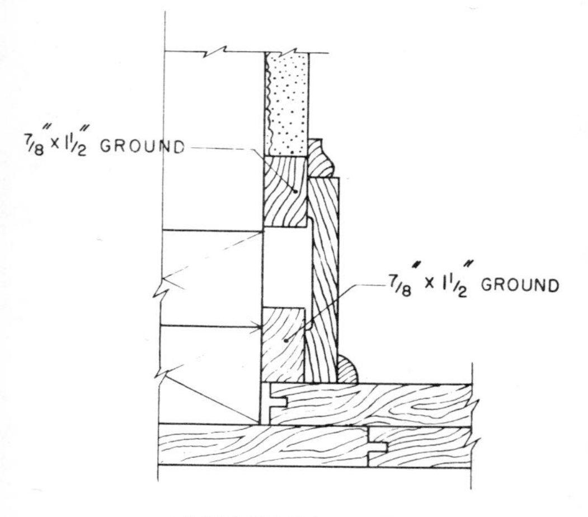

BASEBOARD DETAIL
GROUNDS ON FRAMING MEMBERS

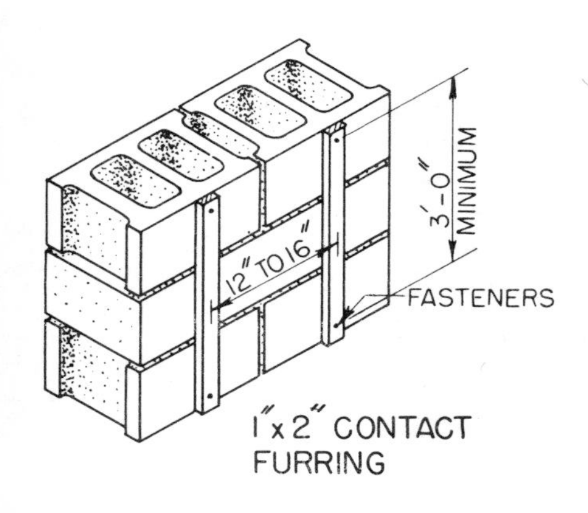

1" x 2" CONTACT FURRING

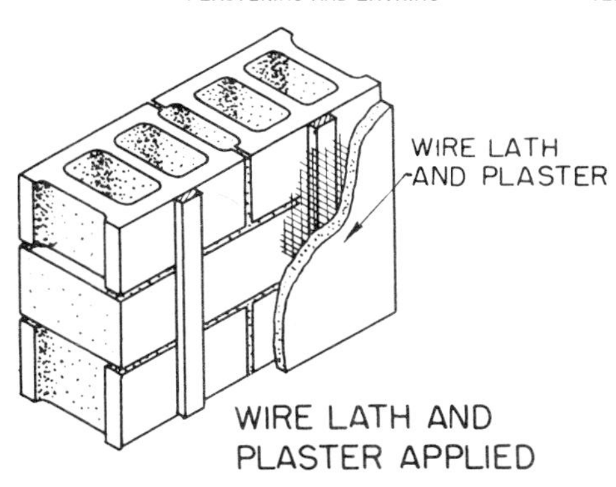

For basement walls below grade it is sometimes desirable to add 1" × 2" cross furring strips on which the lath or other finish material is applied. The extra space between the wall and the finish material will allow moisture, if any, to dry out more readily.

FURRING AND PARTITION TILE

Furring and partition tile are supplied in sizes of 12" × 12" square with thicknesses of 2", 3", 4", 6", 8", 10", and 12". To estimate the quantity required, find the square feet of wall to be furred or partition to be constructed. Deduct for openings, and add an allowance of 2 to 5% for breakage. The number of net square feet represents the number of tile blocks since each block is 12" × 12". The mortar joint sometimes offsets the allowance for breakage.

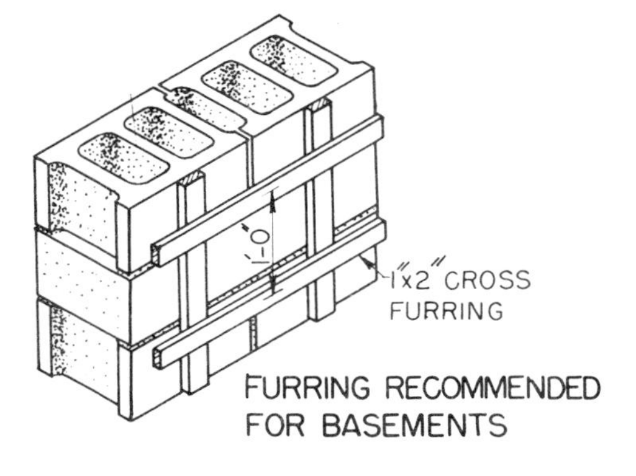

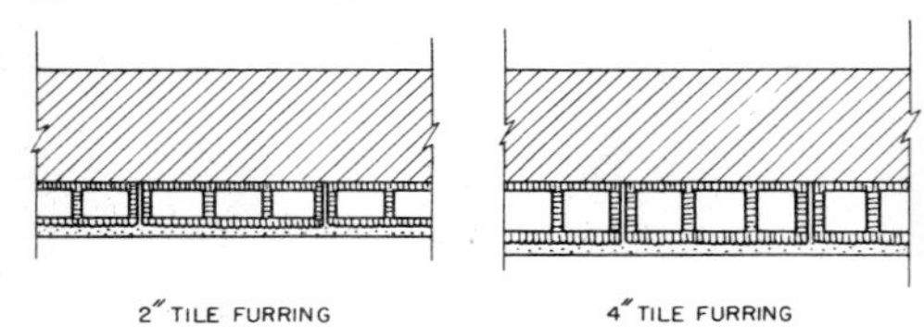

8" EXTERIOR BRICK WALLS WITH HOLLOW TILE FURRING

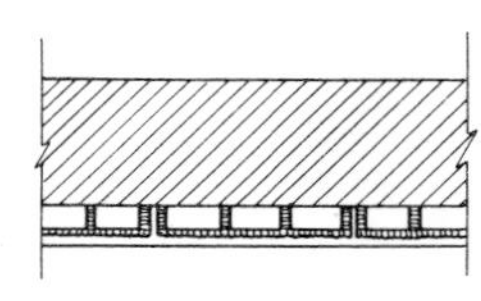

SPLIT FURRING

SIZES AND WEIGHTS OF STANDARD FURRING AND PARTITION TILE

Size of Tile, In.	Weight per Sq. Ft., Lb.	Number Cells	Number Tile per Ton
1½ × 12 × 12	8*	3	250
2 × 12 × 12	9*	3	222
2 × 8 × 12	10†	2	200
2 × 12 × 12	15	3	133
3 × 12 × 12	15	3	133
4 × 12 × 12	16	4	125
5 × 12 × 12	22†	6	91
5 × 12 × 12	19	3	105
6 × 12 × 12	22	3	91
6 × 12 × 12	24†	6	84
8 × 12 × 12	30	6	67
10 × 12 × 12	36	6	56
12 × 12 × 12	40	6	50

*Split furring tile.
†Sizes made in some Eastern factories.

VERMICULITE PLASTER

Vermiculite, a mineral product, will not rot, decay, or burn. It is a lightweight material, weighing about 8 lb. per cu. ft. as compared with 100 lb. per cu. ft. for sand aggregate.

Plaster with Vermiculite aggregate weighs about 30 to 40 lb. per cu. ft., as compared with 100 lb. per cu. ft. where an aggregate sand is used.

Vermiculite is furnished in bags containing about 4 cu. ft.

APPROXIMATE NUMBER OF BRIGHT WIRE NAILS PER POUND

W. & M. Wire Gage	Length, In.																											
	3/16	1/4	3/8	1/2	5/8	3/4	7/8	1	1 1/8	1 1/4	1 1/2	1 3/4	2	2 1/4	2 1/2	2 3/4	3	3 1/2	4	4 1/2	5	6	7	8	9	10	11	12
3/8	......	......	......	......	......	......	......	29	26	23	20	17	15	15	12	11	11	8.9	7.9	7.1	6.4	5.2	4.5	4.0	3.4	3.2	2.9	2.7
5/16	......	......	......	......	......	......	......	43	38	34	29	25	22	20	18	16	15	13	11	10	9	7.6	6.5	5.7	5.0	4.5	4.1	3.8
1	......	......	......	......	......	......	......	47	44	40	34	29	26	23	21	20	18	16	14	12	11	9.3	8.0	7.0	6.3	5.7	5.2	4.7
2	......	......	......	......	......	......	......	60	54	48	41	35	31	28	25	23	21	18	16	14	13	11	9.3	8.1	7.2	6.6	6.1	5.6
3	......	......	......	......	......	......	......	67	60	55	47	41	36	32	29	27	25	21	18	16	15	12	11	9.4	8.3	7.6	7.1	6.6
4	......	......	......	......	......	......	......	81	74	66	55	48	41	37	34	31	29	25	22	20	18	15	13	11	9.8	8.9	8.1	7.5
5	......	......	......	......	......	......	......	90	81	74	61	52	45	41	38	35	32	28	24	22	21	18	16	14	12	12	11	9.6
6	......	......	......	213	174	149	128	113	101	91	76	65	58	52	47	43	39	34	29	26	24	20	18	16	15	13	11	11
7	......	......	......	250	205	174	148	132	120	110	92	78	70	61	55	51	47	40	35	31	28	24	19	18	16	14	13	12
8	......	......	......	272	238	198	174	153	139	126	106	93	82	74	66	61	56	48	42	38	34	28	24	21	19	17	15	14
9	......	......	......	348	286	238	213	185	170	152	128	112	99	87	79	71	67	58	50	45	41	34	29	25	23	21	19	17
10	......	......	......	469	373	320	277	242	216	196	165	142	124	111	100	91	84	71	62	55	49	42	36	31	27	25	23	21
11	......	......	......	510	417	366	323	285	254	233	200	171	149	136	122	111	103	87	77	69	61	52	44	39	35	31	29	26
12	......	......	......	740	603	511	442	397	351	327	268	229	204	182	161	149	137	118	103	95	87	71	63	56	50	45	40	36
13	......	......	1,356	1,017	802	688	590	508	458	412	348	297	260	232	209	190	175	153	138	123	110	93						
14	......	2,293	1,664	1,290	1,037	863	765	667	586	536	459	398	350	312	278	256	233	201	176	157	140	117						
15	......	2,899	2,213	1,519	1,316	1,132	971	869	787	694	578	501	437	390	351	317	290	246	220	196	177	145						
16	......	3,932	2,720	2,142	1,708	1,414	1,229	1,099	973	872	739	635	553	496	452	410	370	318	277	248	226							
17	......	5,316	3,890	2,700	2,306	1,904	1,581	1,409	1,253	1,139	956	831	746	666	590	532	486	418	360	322	295							
18	......	7,520	5,072	3,824	3,130	2,608	2,248	1,976	1,760	1,590	1,338	1,150	996	890	820	740	680	585	507	448	412							
19	......	9,920	6,860	5,075	4,132	3,508	2,816	2,556	2,284	2,096	1,772	1,590	1,390	1,205	1,060	970	895	800										
20	18,620	14,050	9,432	7,164	5,686	4,795	4,230	3,596	3,225	2,893	2,412	2,070	1,810	1,620	1,450	1,315	1,215	1,035										
21	23,260	17,252	12,000	8,920	7,232	6,052	5,272	4,576	4,020	3,640	3,040	2,665	2,310															
22	28,528	21,508	14,676	11,776	9,276	7,672																						
23	35,864	27,039	18,026	13,519	10,815	9,013																						
24	44,936	34,018	22,678	17,008	13,607	11,339																						
25	57,357	43,243	28,828	21,622	17,297	14,414																						

These are approximate average figures for ordinary wire nails. Heavy-gage nails and large heads will run correspondingly fewer per pound; slim nails and small heads correspondingly more.
Galvanized nails, copper, and copper alloys run about 10% fewer per pound; aluminum nails run about three times as many per pound as the count for corresponding sizes of steel nails shown in table above.

VERMICULITE PLASTER PROPORTIONS AND QUANTITIES

Type of Construction	Actual Plaster Thickness	Coat	Recommended Volume Proportions	Required per 100 Sq. Yd.	
				Bags of Plaster	Cu. Ft. Vermiculite
Metal Lath, 5/8" grounds	3/8"	Scratch	1:2	11	22
	1/4"	Brown	1:3	8	24
Metal Lath, 3/4" grounds	3/8"	Scratch	1:2	10	20
	3/8"	Brown	1:3	10	30
Insulation and 3/8" rock lath, 7/8" grounds	1/4"	Scratch	1:2	6	12
	1/4"	Brown	1:3	6	18
2" solid partition (channels), 1 3/4" grounds	3/8"	Scratch (one side only)	1:2	11	22
	1/4"	Brown (one side only)	1:3	7	21
Brick and clay tile, 5/8" grounds	1/4"	Scratch	1:2	5	10
	3/8"	Brown	1:3	6	18

MILLWORK

Millwork is that part of carpentry work which has been prepared by a lumber mill. The estimator figures supply items of millwork from the mill's catalogues, while the contractors figure the same by just adding the percentage to the mill's prices.

It is always good practice for the estimator to make sketches of each piece of millwork listed. A small sketch often gives more information than a lengthy description. In listing windows, the estimator must be certain whether the window dimensions are sash sizes or masonry-opening sizes. In frame openings, windows are usually listed in sash sizes, while it is common practice on the part of the architect to show masonry-opening sizes on his plans for masonry buildings.

There is a considerable difference in the cost of doors and windows depending on the thicknesses of these items. The most common thicknesses are 1⅜" and 1¾". In stating the size of doors or windows, the width is given first, then the height, and finally the thickness.

Exterior trim, for water table, eaves, etc., must be listed carefully to correspond to the work called for on the plans or in the specifications.

Interior trim, such as baseboards, chair rails, wall moldings, and handrails, can usually be listed by stating the number of linear feet and their sizes. Sometimes the trim for doors and windows is included in the price of the door and window.

ITEMS GENERALLY INCLUDED UNDER MILLWORK

STRIP MATERIALS: Baseboard, picture molding, cove molding, chair rail.

OTHER ITEMS: Doors, door frames, screen doors, storm doors, window sash, window frames, screens, inside window,and door trim, cabinets, stairways, fireplace mantels, store fixtures, louvers.

WOOD FLOORING

Wood flooring is usually 25/32 of an inch thick and is tongued and grooved or "matched." The usual widths are 2¼ and 3¼, made from rough stoul 3" and 4". When estimating, use rough dimension.

Since there are numerous grades of flooring, it is important for the estimator to estimate the price for the kind of work specified. Oak flooring may be specified as red oak or white oak in different grades such as clear, select No. 1 or No. 2. This is also true of maple and pine flooring.

There is a considerable loss of surface due to stock requirements and there is also quite some loss or waste in cutting on the job. In estimating, therefor it is essential to allow between one third and one half to the net floor area in order to get the correct number of board feet.

Wood flooring is generally laid on sub-flooring which may be ordinary 1" boards laid diagonally, or panels of plywood.

When estimating wood flooring, find the square feet of surface to be covered then add 20% for 1" × 6" underflooring laid diagonally and about 25% per thousand board feet. A carpenter will take about 28 hours to lay 1000 board feet.

For finish flooring it requires about 14 hours for laying 250 board feet.

PORCH FLOORS—Allow for projection of the flooring beyond the face of the house; add 25% for 4" flooring and 20% for 6" flooring. A carpenter will lay about 315 board feet of porch flooring within a 15-hour period.

DOORS—Exterior doors will take about 7 to 9 carpenter hours; basement door 3½ hours; interior doors 4 hours; combination doors about 3½ hours; mirror and french doors about 10 hours.

CABINETS—Receive the same finishes and materials as doors and windows.

KITCHEN CABINETS—About 6 carpenter hours

BOOK CASES—6 carpenter hours

MANTEL—Price varies greatly. Depending on size, and quality.

FRONT ENTRANCE—List the style and catalogue number. It takes about six hours to install front entrance door.

CORNER BOARDS—List the number of pieces required, and determine the number of board feet. Add 25% for 1¼" stock; 50% for 1½" stock. It takes about 65 labor hours to install per thousand board feet.

CORNICE—Total length of corner board multiplied by total width including width of wall moulding material, frieze, fascia, etc. Length of cornice × width = sq. ft. It requires about 24 labor hours per hundred linear feet.

BASE AND WATER TABLE—Get total linear feet and multiply by width to get sq. ft. then into board feet; add 12%.

PORCH STEPS—Determine board feet for stringers, risers, treads, and linear feet of mouldings.

Storm windows and screens are made at the mill and one carpenter will fit two screens or two storm windows per hour.

Base board is figured in linear feet and one carpenter will install ai ·ut 16 linear feet in one hour.

PICTURE MOULDING—A carpenter will install about 24 linear feet per hour.

MAIN STAIRWAY—No definite estimate can be made due to the many variations. It takes about 14 hours to install a stairway.

BASEMENT STAIR—It will take about 10 hours for a carpenter to install.

WINDOWS—Basement sash can be installed in about one hour; a double hung window 4 hrs; a casement window about 3½ hrs; a fixed sash in 1¾ to 2½ hours.

PLUMBING

The expression "Mechanical Trades" usually applies to Plumbing, Heating, Electrical and Elevator work. Practically without exception Plumbing and Heating are sublet either by the general contractor or by the owner outside of the general contract. While the general contractor thus has nothing to do with these lines of work, it still will be in his interest and in the interest of the progress of the work to check into these contracts and to make sure that he will work hand in hand with such plumbing and heating contractors. The excavator of the general contractor will have to handle the excavation work needed for these mechanical trades as well as their back filling, street cutting and patching. Some mechanical trade contractors will agree to handle these sections of the work but it will prove more economical to have all excavating work digging of trenches and so on, done by one and the same contractor.

It is up to the plumbing contractor to furnish and to install all bathtubs, lavatories, sinks and laundry tubs. Plumbing also includes hot water storage tanks and hot water heaters. The installation of all water supply and drainage lines, of all temporary water drainage facilities during the construction of the job, all rough work and all catchbasins and manholes required are part of the plumbing job. Sometimes toilet and bathroom extras like soap and paper holders, hand towel bars, mirrors, and cabinets are specified under plumbing work.

Plumbing work is usually estimated by the quantity survey method which includes listing of all materials. An approximate estimate of plumbing costs for the average house can be easily arrived at by listing all fixtures, then estimating the cost of the total job by multiplying the number of fixtures by the cost per fixture installed, which would include the cost of all fixtures, piping, faucets, labor, and so on.

NAILS REQUIRED FOR OUTSIDE TRIM

Item	Dimensions, In.	Number of Nailings	Common Wire Nails No. of Lbs per 1000 Bd. Ft.	
			8d	10d
Base and water table	7/8 × 3			
As a unit	7/8 × 6	4	43	
Base and water table	7/8 × 3			
As a unit	7/8 × 7	4	38.5	
Base and water table	1¼ × 3½			
As a unit	1¼ × 6	4	......	49
Base and water table	1¼ × 3½			
As a unit	1¼ × 7	4	......	43
Corner boards	7/8 × 3			
As a unit	7/8 × 4	5	69	
Corner boards	7/8 × 4			
As a unit	7/8 × 5	5	53.5	

HEATING

The main problem in estimating the cost of heating is the correct determining of heat losses. Some heating contractors use rule of thumb methods, but these rule of thumb methods may prove very detrimental and may in an inadequate installation cause future loss of business.

Heat losses should be expressed in B.T.U. (British Thermal Unit). One B.T.U. is the amount of heat required to raise the temperature of one pound of water 1°F.

The following types of heating systems are in common use:

One Pipe Steam System

Where Used

1. Residential Structures
2. Apartment Buildings
3. Stores
4. Factories
5. Small Public Buildings
6. Theatres

Advantages

1. Low cost of material, labor and maintenance.
2. Easy to install.
3. Very few "Steam Specialties" required.
4. Existing systems can easily be converted into one pipe vapor system by addition of vacuum air valves in place of air vents.

Disadvantages

1. Steam and water flow in opposite directions in branches and risers, necessitating large size pipes.
2. Water hammer sometimes prevalent.
3. Air vented directly into room. (odor)
4. Considerable time and pressure required to vent air from each radiator before heating becomes effective, thereby wasting fuel.
5. Not suitable for use with modern "convector" type radiators.
6. This system doomed to become obsolete.

Vacuum System

Where Used

1. Office Buildings
2. Hospitals
3. Schools
4. Apartment Buildings
5. Central Station Heating Plants for Housing Developments
6. Hotels
7. Department Stores
8. Public Buildings

Advantages

1. Steam circulated at low pressure (under 5 lbs.) due to constant vacuum in return mains caused by vacuum pump.
2. Radiator temperature may be "modulated" within limited range.
3. Suitable for automatic control.
4. Radiators may be placed below water line, if pump is still lower.

Disadvantages

1. High cost due to expensive steam specialties and vacuum pump.
2. Frequent maintenance required on large installation.

Two Pipe Vapor System

Where Used	Advantages	Disadvantages
1. Residential Structures 2. Small Apartment Buildings 3. Small Public Buildings 4. Stores	1. Extremely low pressure (1 to 2 oz.) and even partial vacuum sufficient to circulate steam throughout system (because no air can accumulate in radiators due to thermostatic traps). 2. Modulating type steam valves may be used, thus making it possible to "throttle" the radiator. (Not possible with one pipe system.) 3. No water hammer or odor. Quick heating-up period. Very efficient. 4. System ideal in response to automatic control. 5. May be used with modern "convectors."	1. Higher cost than one pipe system. Expensive valves, traps and other steam specialties required. 2. More labor required for installation than one pipe system.

Radiant Heating

Where Used	Advantages	Disadvantages
1. Any type of Structure	1. Healthful heat—no stratification of room temperature. (3 or 4 degrees) 2. All room surfaces become warm—comfort in any part of the room. 3. Lower temperatures can be maintained with equal comfort. 4. No heating devices visible. 5. Economical to operate—shows savings over conventional systems.	1. Not suitable for addition of summer cooling. Complete new system must be installed for air conditioning (unless warm air is used). 2. Costly to install in existing buildings.

Warm Air System

Where Used	Advantages	Disadvantages
1. Residential Structures 2. Small Public Buildings	1. This system provides all the requirements of winter air conditioning, via: heating, humidification, air cleaning and circulation. 2. May be converted into year 'round air conditioning system by addition of summer cooling equipment. Same duct work can be used. 3. System can't freeze in winter. 4. Can circulate all outdoor air in fall and spring.	1. Cold floor line temperatures likely if distribution not planned with care. 2. Ductwork sometimes requires additional headroom in basement.

Hot Water System

Where Used	Advantages	Disadvantages
1. Residential Structures 2. Apartment Buildings 3. Stores 4. Factories 5. Small Public Buildings 6. Housing Developments 7. Hospitals	1. Extremely efficient due to instant circulation of water. 2. Low temperature heat for mild weather. 3. Radiators "hold" heat for a long time.	1. System always full of water—must be drained when not in use in cold climates. 2. May overheat when weather turns mild suddenly.

NAILS REQUIRED FOR DOOR AND WINDOW FRAMES, INSIDE DOOR AND WINDOW TRIM, INSIDE BASE, AND PICTURE MOLDING

Number of Items Taken in a Unit. Unit = 100 Lin. Ft.	Kind of Nail	Finish Nails			
		4d	6d	8d	10d
5 door or window frames	Common	......	....	....	1
5 inside door jambs	Casing	......	....	....	1
10 inside door trim including stops and casing saddles	Casing	.25	.25	1.25	
Inside base, 2-member per 100 lin. ft.	Finish	......	.55		
	Casing	......	....	1.25	
5 inside window trim including stops	Casing	.75	1.25	1.25	
Inside base, 3-member per 100 lin. ft.	Finish	......	.82		
	Casing	......	....	1.25	
Picture molding per 100 lin. ft.	Finish	......	.5		
	Casing	......	....	.6	

ALUMINUM SLIDING GLASS DOORS

Glass Size	Frame Size Width–Height	Rough Opening
33″ x 76¾″	6′ x 6′-10¾″	6′-0½″ x 6′-11¼″
45″ x 76¾″	8′ x 6′-10¾″	8′-0½″ x 6′-11¼″
57″ x 76¾″	10′ x 6′-10¾″	10′-0½″ x 6′-11¼″
33″ x 76¾″	9′ x 6′-10¾″	9′-0½″ x 6′-11¼″
45″ x 76¾″	12′ x 6′-10¾″	12′-0½″ x 6′-11¼″
57″ x 76¾″	15′ x 6′-10¾″	15′-0½″ x 6′-11¼″
33″ x 76¾″	11′-11″ x 6′-10¼″	11′-11½″ x 6′-11¼″
45″ x 76¾″	15′-11″ x 6′-10¼″	15′-11½″ x 6′-11¼″
57″ x 76¾″	19′-11″ x 6′-10¼″	19′-11½″ x 6′-11¼″

x = Sliding Panel
o = Fixed Panel
Note: Glass Door illustrated is "outside looking in"

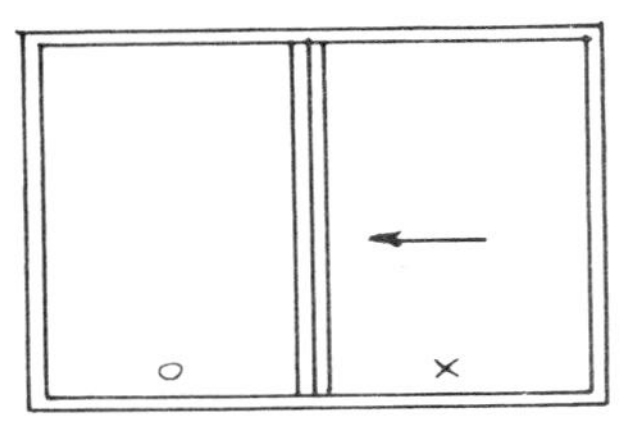

EXTRUDED ALUMINUM STOCK DOOR AND FRAME SIZES

Masonry Opening Size	Frame Size	Door Opg. Size	Actual Door Size
3′-4″ x 7′-2″	3′-3½″ x 7′-1¾″	3′-0″ x 7′-0″	35 13/16″ x 83¼″
3′-10″ x 7′-2″	3′-9½″ x 7′-1¾″	3′-6″ x 7′-0″	41 13/16″ x 83¼″
5′-4″ x 7′-2″	5′-3½″ x 7′-1¾″	5′-0″ x 7′-0″	29 13/16″ x 83¼″

PAINTING

The most accurate and perhaps the best method of estimating painting is to find the actual surface area to be painted from the plans and specifications or to take actual field measurements. As in everything else, painting costs are based on two major items, namely, labor and material. The cost of labor depends on the present-day labor scales, and the cost of material depends on the grade, or quality, and quantity of paint used. Estimating the quantities of material by finding the surface areas to be painted is relatively simple, but the labor quantities present a much more difficult problem.

For example: A plain wall surface may have the same surface area as a cornice, but it takes much longer to paint the cornice owing to its height, which requires the erection of scaffolding. Care must be taken in pricing any piece of work, as conditions on each job are different. The following table, giving methods of measuring and listing painting quantities, is based on the actual performance of numberous jobs and represents a fairly accurate method of establishing quantities.

METHODS OF MEASURING AND LISTING PAINTING QUANTITIES

Description of Item	Unit of Measure	'Multiply by Factor
Clapboards or drop siding (no deductions for areas less than 10' × 10')	Find actual area	Add 10% to surface
Shingle siding (no deductions for areas less than 10' × 10')	Find actual area	1½ times area
Brick, wood, stucco, cement, stone walls (no deductions for areas less than 10' × 10')	Find actual area	
Eaves—plain, painted same color as side walls	Find actual area	1½ times area
Eaves—different color from side walls	Find actual area	2 times area
Eaves—with rafters running through	Find actual area	3 times area
Eaves—over 20' above ground	Find actual area	Add ½ of area to each 10' of height
Eaves—over brick, stucco, or stone walls	Find actual area	3 times area
Exterior cornices:		
Plain	Find actual area	2 times area
Fancy	Find actual area	3 times area
Down spouts and gutters:		
Plain	Find actual area	2 times area
Fancy	Find actual area	3 times area
Blinds and shutters:		
Plain	Area of outer faces	2 times area
Slatted	Area of outer faces	4 times area
Columns and pilasters:		
Plain	Area, sq. ft.	
Fluted	Area, sq. ft.	1½ times area
Paneled	Area, sq. ft.	2 times area

METHODS OF MEASURING AND LISTING PAINTING QUANTITIES
(Continued)

Description of Item	Unit of Measure	Multiply by Factor
Moldings:		
If under 12" in girth	Figure 1 sq. ft. per lin. ft.	
If over 12" in girth	Take actual area	
Exterior doors and frames:		
Figure no door less than 3' × 7', Allow for frame, add 2' to width and 1' to height	Figure all doors 40 sq. ft.	2 times area for both sides
Containing small lights of glass	Add 2 sq. ft. for each additional light	2 times area for both sides
Door frames only (no door)	Allow area of opening for both sides	
Exterior windows:		
Figure no window less than 3' × 6'; add 2' to both width and height of opening	Figure all windows 40 sq. ft.	
Sash containing more than one light	Add 2 sq. ft. for each additional light	
Interior doors, jambs, and casings: Figure no door less than 3' × 7', allow for frame, add 2' to width and 1' to height	Figure all doors 40 sq. ft.	Do not deduct for glass in doors
Containing small lights of glass	Add 2 sq. ft. for each additional light	2 times area for both sides of door
Stairs:		
Add 2' to length of treads and risers to allow for stair strings. Figure 2' width of tread and riser	Multiply width (2') by length of thred. Then multiply by Number of treads, for total sq. ft.	
Wood ceilings	Find actual area	No deductions for openings less than 10' × 10'
Floors	Actual area	
Plastered walls and ceilings	Actual area of wall and ceiling, sq. ft.	Do not deduct for door and window openings
Radiators	For each front foot, multiply face area by 7	
Bookcases, cupboards, closets	Find area of front	Multiply area (front) by 3
Wainscoting:		
Plain	Find actual area	
Paneled	Find actual area	Multiply area by 2

MATERIAL AND LABOR REQUIRED
FOR ONE COAT OF PAINT

Material	Sq. Ft. per Gal.	Sq. Ft. per Hr.
Paint, lead and oil, outside	400 to 450	150 to 200
Paint, lead and oil, walls	450	200
Paint, lead and oil, trim	450	175
Ready-mixed paint	450	
Aluminum paint	600	250
Floor paint	400	300
Cement paint on rough concrete	150 to 400	125 to 300
Calcimine	300 or 2 lb.	250
Flat paint on walls	450	175
Flat paint on trim	450	150
Enamel on trim	400	125
Enamel on walls	400	150
Red lead on wood	350	250
Red lead on metal	700	125 to 150
Filler on trim	300	175
Filler on floors	300	250
Varnish on floors	400	300
Varnish on trim	400	150
Varnish remover	150	
Shellac on floors	400	300
Shellac on trim	400	175
Shingle stain (dip)	40	
Shingle stain (brushed)	160	150
Shingle stain (creosote)	200	150
Varnish stain	400	200
Glue size	1,000	1,000
Glazing liquid	500	400
Linseed oil (boiled)	400	400
Turpentine tint	250	
Undercoat	450	175
Plastic paint	75 sq. yd.	
Window frames and sash		4

ESTIMATING WALLPAPER

A roll of wallpaper contains 36 sq. ft. The standard size roll measures 18" wide and 24'-0" long, or 1'-6" × 24'-0" = 36 sq. ft., or 4 sq. yd. A double roll of wallpaper measures 18" wide by 48'-0" long, or 1'-6" × 48'-0" = 72 sq. ft., or 8 sq. yd.

To find the number of rolls of wallpaper required for a room, find the total area of the wall by multiplying the perimeter of the room by the height of the room. The result will be the total number of square feet of surface to be covered. If single rolls are to be used, divide the total number of square feet by 36. If double rolls are to be used, divide the total square feet by 72.

It is customary to allow between 15 and 20% for waste and matching. All window, door, or other openings should be deducted.

EXAMPLE: A room measuring 12' × 16' has a ceiling height of 8'-6". There is one door 2'-6" × 7'-0 , and one window measuring 2'-6" × 4'-6". How many single rolls are required?

SOLUTION: The perimeter of 4 walls is

12 + 16 + 12 + 16 = 56'

56' × 8' (height of wall less 6" baseboard) = 448 sq. ft.

Door area: 2.5 × 7 = 17.5 sq. ft.

Window area: 2.5 × 4.5 = 11.25 sq. ft.

28.75 sq. ft. total area of door and window.

Therefore,

448.00 sq. ft.

− 28.75 sq. ft.

419.25 or 420 sq. ft.

Allow 15% for waste and matching.

.15 × 420 = 63 sq. ft.

Therefore,

420 sq. ft.

+ 63

483 net sq. ft. of paper required

483 ÷ 36 = 13.44 or 14 single rolls.

PASTE REQUIREMENTS FOR WALLPAPER

Weight of Paper	Gal. of Paste	Number of Rolls	Number of Paste Coats
Light- or medium-weight paper	1	12 single rolls	1
Heavy or rough-textured paper	1	4 to 6 single rolls	2 to 3 coats

An experienced paper hanger should fit and hang between 12 and 16 single rolls of wallpaper per 8-hr. day. This is for light- or medium-weight paper. Where a good grade of medium or heavy paper is used and a first-class job is desired, a paper hanger can fit and hang between 10 and 12 single rolls per 8-hr. day.

MENSURATION

SQUARE MEASURE

144 square inches = 1 square foot
9 square feet = 1 square yard
100 square feet = 1 square
30¼ square yards = 1 square rod
160 square rods = 1 acre
43,560 square feet = 4,840 square yards
640 acres = 1 square mile

CUBIC MEASURE

1,728 cubic inches = 1 cubic foot
27 cubic feet = 1 cubic yard
128 cubic feet = 1 cord
24¾ cubic feet = 1 perch

LINEAR MEASURE

12 inches = 1 foot
3 feet = 1 yard
5½ yards = 16½ feet = 1 rod, pole, or perch
40 poles = 220 yards = 1 furlong
8 furlongs = 1,760 yards = 1 mile
3 miles (U.S. nautical) = 1 league
4 inches = 1 hand
9 inches = 1 span

SURVEYOR'S MEASURE

7.92 inches = 1 link
100 links = 66 feet = 4 rods = 1 chain
80 chains = 1 mile
33⅓ inches = 2.75 feet = 1 vara

INCHES INTO DECIMALS OF A FOOT

Inches

Inch	0	1	2	3	4	5	6	7	8	9	10	11
Decimals of a Foot												
		.083	.167	.250	.333	.4167	.500	.583	.667	.750	.833	.916
1/16	.005	.088	.171	.255	.338	.421	.505	.588	.671	.755	.838	.921
1/8	.010	.093	.177	.260	.343	.427	.510	.593	.677	.760	.843	.927
3/16	.015	.099	.182	.265	.349	.432	.515	.599	.682	.765	.849	.932
1/4	.020	.104	.187	.270	.354	.437	.520	.604	.687	.770	.854	.937
5/16	.026	.109	.192	.276	.359	.442	.526	.609	.692	.776	.859	.942
3/8	.031	.114	.197	.281	.364	.447	.531	.614	.697	.781	.864	.947
7/16	.036	.119	.203	.286	.369	.453	.536	.619	.703	.786	.869	.953
1/2	.041	.125	.208	.291	.375	.458	.541	.625	.708	.791	.875	.958
9/16	.046	,130	.213	.296	.380	.463	.546	.630	.713	.796	.880	.963
5/8	.052	.135	.218	.302	.385	.468	.552	.635	.718	.802	.885	.968
11/16	.057	.140	.224	.307	.390	.474	.557	.640	.724	.807	.890	.974
3/4	.062	.145	.229	.312	.395	.479	.562	.645	.729	.812	.895	.979
13/16	.067	.151	.234	.317	.401	.484	.567	.651	.734	.817	.901	.984
7/8	.072	.156	.239	.322	.406	.489	.572	.656	.739	.822	.906	.989
29/32	.075	.158	.242	.325	.408	.492	.575	.658	.742	.825	.908	.992
15/16	.078	.161	.244	.328	.411	.494	.578	.661	.744	.828	.911	.994
31/32	.080	.164	.247	.330	.414	.497	.580	.664	.747	.830	.914	.997

FINDING THE NUMBER OF SQUARE YARDS IN WALLS AND CEILINGS OF VARIOUS ROOM SIZES

The following tables give the number of square yards of walls and ceiling in rooms ranging from 3'-0" × 3'-0", the smallest size, to 24'-0" × 22'-0", the largest size, and with ceiling heights from 7'-0" to 12'-0" in increments of 6".

To find the number of square yards of wall and ceiling in a room measuring 16'-0" × 18'-0" with a ceiling height of 7'-6", proceed as follows: Turn to the table for 7'-6" ceiling height. Find the width of the room, 16', along the left border under the heading marked "Room Size," and follow horizontally across and read the figure under the length of the room, which is 18'-0". The number of square yards thus found is 88.6.

For example: To qualify this figure, note the following:

Room size = 16 × 18 × 7'-6"
16 + 16 + 18 + 18 = 68' the perimeter of room
68 × 7.5 = 510 sq. ft. of walls
16 × 18 = 288 sq. ft. of ceiling area
510 + 288 = 798 sq. ft. of ceiling and walls
798 ÷ 9 = 88.6 sq. yd.

If after finding the number of square yards of wall and ceiling in a room it is desired to know the number of square feet, multiply the square yards found in the table by 9. In a room of 12' × 14', ceiling height 7'/0", there are 59.1 sq. yd. 59.1 × 9 = 631.9 or 632 sq. ft.

These tables may be used in quickly determining areas when estimating for painting, plaster, wallpaper, metal and sheet lath, etc.

CONVERSIONS

To Change	To	Multiply by
Inches	Feet	0.0833
Inches	Millimeters	25.4
Feet	Inches	12
Feet	Yards	0.3333
Yards	Feet	3
Square Inches	Square Feet	0.00694
Square Feet	Square Inches	144
Square Feet	Square Yards	0.1111
Square Yards	Square Feet	9
Cubic Inches	Cubic Feet	0.00058
Cubic Feet	Cubic Inches	1728
Cubic Feet	Cubic Yards	0.03703
Cubic Yards	Cubic Feet	27
Cubic Inches	Gallons	0.00433
Cubic Feet	Gallons	7.48
Gallons	Cubic Inches	231
Gallons	Cubic Feet	0.1337
Gallons	Pounds of Water	8.33
Pounds of Water	Gallons	0.12004
Ounces	Pounds	0.0625
Pounds	Ounces	16

NUMBER OF SQUARE YARDS OF WALLS AND CEILINGS IN ROOMS WITH 7'-0" CEILING HEIGHT

Room Size	3	4	5	6	7	8	9	10	11	12	13	14	15	16	17	18	19	20	21	22
3	10.3	12.2	14.1	16.0	17.8	19.7	21.6	23.5	25.4	27.3	29.2	31.1	33.0	34.8	36.7	38.6	40.5	42.4	44.3	46.2
4	12.2	14.2	16.2	18.2	20.2	22.2	24.2	26.2	28.2	30.2	32.2	34.2	36.2	38.2	40.2	42.2	44.2	46.2	48.2	50.2
5	14.1	16.2	18.3	20.4	22.5	24.6	26.7	28.8	31.0	33.1	35.2	37.3	39.4	41.5	43.6	45.7	47.8	50.0	52.1	54.2
6	16.0	18.2	20.4	22.6	24.8	27.1	29.3	31.5	33.7	36.0	38.2	40.4	42.6	44.8	47.1	49.3	51.5	53.7	55.8	58.2
7	17.8	20.2	22.5	24.8	27.2	29.5	31.8	34.2	36.5	38.8	41.2	43.5	45.8	48.2	50.5	52.8	55.2	57.5	59.8	62.2
8	19.7	22.2	24.6	27.1	29.5	32.0	34.4	36.8	39.3	41.7	44.2	46.6	49.1	51.5	54.0	56.4	58.8	61.3	63.7	66.2
9	21.6	24.2	26.7	29.3	31.8	34.4	37.0	39.5	42.1	44.6	47.2	49.7	52.3	54.8	57.4	60.0	62.5	65.1	67.6	70.2
10	23.5	26.2	28.8	31.5	34.2	36.8	39.5	42.2	44.8	47.5	50.2	52.8	55.5	58.2	60.8	63.5	66.2	68.8	71.5	74.2
11	25.4	28.2	31.0	33.7	36.5	39.3	42.1	44.8	47.6	50.4	53.2	56.0	58.7	61.5	64.3	67.1	69.8	72.6	75.4	78.2
12	27.3	30.2	33.1	36.0	38.8	41.7	44.6	47.5	50.4	53.3	56.2	59.1	62.0	64.8	67.7	70.6	73.5	76.4	79.3	82.2
13	29.2	32.2	35.2	38.2	41.2	42.2	47.2	50.2	53.2	56.2	59.2	62.2	65.2	68.2	71.2	74.2	77.2	80.2	83.2	86.2
14	31.1	34.2	37.3	40.4	43.5	46.6	49.7	52.8	56.0	59.1	62.2	65.3	68.4	71.5	74.6	77.7	80.8	84.0	87.1	90.2
15	33.0	36.2	39.4	42.6	45.8	49.1	52.3	55.5	58.7	62.0	65.2	68.4	71.6	74.8	78.1	81.3	84.5	87.7	91.0	94.2
16	34.8	38.2	41.5	44.8	48.2	51.5	54.8	58.2	61.5	64.8	68.2	71.5	74.8	78.2	81.5	84.8	88.2	91.5	94.8	98.2
17	36.7	40.2	43.6	47.1	50.5	54.0	57.4	60.8	64.3	67.7	71.2	74.6	78.1	81.5	85.5	88.4	91.8	95.3	98.7	102.2
18	38.6	42.2	45.7	49.3	52.8	56.4	60.0	63.5	67.1	70.6	74.2	77.7	81.3	84.8	88.4	92.0	95.5	99.1	102.6	106.2
19	40.5	44.2	47.8	51.5	55.2	58.8	62.5	66.2	69.8	73.5	77.2	80.8	84.5	88.2	91.8	95.5	99.2	102.8	106.5	110.2
20	42.4	46.2	50.0	53.7	57.5	61.3	65.1	68.8	72.6	76.4	80.2	84.0	87.7	91.5	95.3	99.1	102.8	106.6	110.4	114.0
21	44.3	48.2	52.1	55.8	59.8	63.7	67.6	71.5	75.4	79.3	83.2	87.1	91.0	94.8	98.7	102.6	106.5	110.4	114.3	118.2
22	46.2	50.2	54.2	58.2	62.2	66.2	70.2	74.2	78.2	82.2	86.2	90.2	94.2	98.2	102.2	106.2	110.2	114.2	118.2	122.2
23	48.0	52.2	56.3	60.4	64.5	68.6	72.1	76.8	81.0	85.1	89.2	93.3	97.4	101.5	105.6	109.7	113.8	118.0	122.1	126.2
24	50.0	54.2	58.4	62.6	66.8	71.1	75.3	79.5	83.7	88.0	92.2	96.4	100.6	104.8	109.1	113.3	117.5	121.7	126.0	130.2

7'-6"

NUMBER OF SQUARE YARDS OF WALLS AND CEILINGS IN ROOMS WITH 7'-6" CEILING HEIGHT

Room Size	3	4	5	6	7	8	9	10	11	12	13	14	15	16	17	18	19	20	21	22
3	11.0	13.0	15.0	17.0	19.0	21.0	23.0	25.0	27.0	29.0	31.0	33.0	35.0	37.0	39.0	41.0	43.0	45.0	47.0	49.0
4	13.0	15.1	17.2	19.3	21.4	23.5	25.6	27.7	29.8	32.0	34.1	36.2	38.3	40.4	42.5	44.0	46.7	49.8	51.0	53.1
5	15.0	17.2	19.4	21.6	23.8	26.1	28.3	30.5	32.7	35.0	37.2	39.4	41.6	43.8	46.1	48.3	50.5	52.7	55.0	57.2
6	17.0	19.3	21.6	24.0	26.3	28.6	31.0	33.3	35.6	38.0	40.3	42.6	45.0	47.3	49.6	52.0	54.3	56.6	59.0	61.3
7	19.0	21.4	23.8	26.3	28.7	31.2	33.6	36.1	38.5	41.0	43.4	45.8	48.3	50.7	53.2	55.6	58.1	60.5	63.0	65.4
8	21.0	23.5	26.1	28.6	31.2	33.7	36.3	38.8	41.4	44.0	46.5	49.1	51.6	54.2	56.7	59.3	61.8	64.4	67.0	69.5
9	23.0	25.6	28.3	31.0	33.6	36.3	39.0	41.6	44.3	47.0	49.6	52.3	55.0	57.6	60.3	63.0	65.6	67.3	71.0	73.6
10	25.0	27.7	30.5	33.0	36.1	38.8	41.6	44.4	47.2	50.0	52.7	55.5	58.3	61.1	63.8	66.6	69.4	72.7	75.0	77.7
11	27.0	29.8	32.7	35.6	38.5	41.4	44.3	47.2	50.1	53.0	55.8	58.7	61.6	64.5	67.4	70.3	73.2	76.1	79.0	81.8
12	29.0	32.0	35.0	38.0	41.0	44.0	47.0	50.0	53.0	56.0	59.0	62.0	65.0	68.0	71.0	74.0	77.0	80.0	83.0	86.0
13	31.0	34.1	37.2	40.3	43.0	46.5	49.6	52.7	55.8	59.0	62.1	65.2	68.3	71.4	74.5	77.6	80.7	84.8	87.0	90.1
14	33.0	36.2	39.4	42.6	45.8	49.1	52.3	55.5	58.7	62.0	65.2	68.4	71.6	74.8	78.1	81.3	84.5	87.7	91.0	94.2
15	35.0	38.3	41.6	45.0	48.3	51.6	55.0	58.3	61.6	65.0	68.3	71.6	75.0	78.3	81.6	85.0	88.3	91.6	95.0	98.3
16	37.0	40.4	43.8	47.3	50.7	54.2	57.6	61.1	64.5	68.0	71.4	74.8	78.3	81.7	85.2	88.6	92.1	95.5	99.0	102.4
17	39.0	42.5	46.1	49.6	53.2	56.7	60.3	63.8	67.4	71.0	74.5	78.1	81.6	85.2	88.7	92.3	95.8	99.4	103.0	106.5
18	41.0	44.6	48.3	52.0	55.6	59.3	63.0	66.6	70.3	74.0	77.6	81.2	85.0	88.6	92.3	96.0	99.6	103.3	107.0	110.6
19	43.0	46.7	50.5	54.3	58.1	61.8	65.6	69.4	73.2	77.0	80.7	84.5	88.3	92.1	95.8	99.6	103.4	107.2	111.0	114.7
20	45.0	49.8	52.7	56.6	60.5	64.4	68.3	72.2	76.1	80.0	84.8	87.7	91.6	95.5	99.4	103.3	107.2	111.1	115.0	118.8
21	47.0	51.0	55.0	59.0	63.0	67.0	71.0	75.0	79.0	83.0	87.0	91.0	95.0	99.0	103.0	107.0	111.0	115.0	119.0	123.0
22	49.0	53.1	57.2	61.3	65.4	69.5	73.6	77.7	81.8	86.0	90.1	94.2	98.3	102.4	106.5	111.6	114.7	118.8	123.0	127.0
23	51.0	55.2	59.4	63.6	67.8	72.1	76.3	80.5	84.7	89.0	93.2	97.4	101.6	105.8	110.1	114.3	118.5	122.7	127.0	131.2
24	53.0	57.3	61.6	66.0	70.3	74.6	79.3	83.3	87.7	92.0	96.3	100.7	105.0	109.3	113.6	118.0	122.3	126.6	131.0	135.3

NUMBER OF SQUARE YARDS OF WALLS AND CEILING IN ROOMS WITH 8'-0" CEILING HEIGHT

Room Size	3	4	5	6	7	8	9	10	11	12	13	14	15	16	17	18	19	20	21	22
3	11.6	13.7	15.8	18.0	20.1	22.2	24.3	26.4	28.5	30.6	32.7	34.8	37.0	39.1	41.2	43.3	45.4	47.5	49.6	51.7
4	13.7	16.0	18.2	20.4	22.6	24.8	27.1	29.3	31.5	33.7	36.0	38.2	40.4	42.6	44.8	47.1	49.3	51.5	53.7	56.0
5	15.8	18.2	20.5	22.8	25.2	27.5	29.8	32.2	34.5	36.8	39.2	41.5	43.8	46.2	48.5	50.8	53.2	55.5	57.8	60.2
6	18.0	20.4	22.8	25.3	27.7	30.2	32.6	35.1	37.5	40.0	42.4	44.8	47.3	49.7	52.2	54.6	57.1	59.5	62.0	64.4
7	20.1	22.6	25.2	27.7	30.3	32.8	35.4	38.0	40.5	43.1	45.6	48.2	50.7	53.3	55.8	58.4	61.0	63.5	66.1	68.9
8	22.2	24.8	27.5	30.2	32.8	35.5	38.2	40.8	43.5	46.2	48.8	51.5	54.2	56.8	59.5	62.2	64.8	67.5	70.2	72.8
9	24.3	27.1	29.8	32.6	35.4	38.2	41.0	43.7	46.5	49.3	52.1	54.8	57.6	60.4	63.2	66.0	68.7	71.5	74.3	77.1
10	26.4	29.3	32.2	35.1	38.0	40.8	43.7	46.6	49.5	52.4	55.3	58.2	61.1	64.0	66.8	69.7	72.6	75.5	78.4	81.3
11	28.5	31.5	34.5	37.5	40.5	43.5	46.5	49.5	52.5	55.5	58.5	61.5	64.5	67.5	70.5	73.5	76.5	79.5	82.5	85.5
12	30.6	33.7	36.8	40.0	43.1	46.2	49.3	52.4	55.5	58.6	61.7	64.8	68.0	71.1	74.2	77.3	80.4	83.5	86.6	89.7
13	32.7	36.0	39.2	42.4	45.6	48.8	52.1	55.3	58.5	61.7	65.0	68.2	71.4	73.6	77.8	81.1	84.3	87.5	90.7	94.0
14	34.8	38.2	41.5	44.8	48.2	51.5	54.8	58.2	61.5	64.8	68.2	71.5	74.8	78.2	81.5	84.8	88.2	91.5	94.8	98.2
15	37.0	40.4	43.8	47.3	50.7	54.2	57.6	61.1	64.5	68.0	71.4	74.8	78.3	81.7	85.2	88.6	92.1	95.5	99.0	102.4
16	39.1	42.6	46.2	49.7	53.3	56.8	60.4	64.0	67.5	71.1	74.6	78.2	81.7	85.3	88.8	92.4	96.0	99.5	103.1	106.6
17	41.2	44.8	48.5	52.5	55.8	59.5	63.2	66.8	70.5	74.2	77.8	81.5	85.2	88.8	92.5	96.2	99.8	103.5	107.2	110.8
18	43.3	47.1	50.8	54.6	58.4	62.2	66.0	69.7	73.5	77.3	81.1	84.8	88.6	92.4	96.2	100.0	103.7	107.5	111.3	115.1
19	45.4	49.3	53.2	57.1	61.0	64.8	68.7	72.6	76.5	80.4	84.3	88.2	92.1	96.0	99.8	103.7	107.6	111.5	115.4	119.3
20	47.5	51.5	55.5	59.5	63.5	67.5	71.5	75.5	79.5	83.5	87.5	91.5	95.5	99.5	103.5	107.5	111.5	115.5	119.5	123.5
21	49.6	53.7	57.8	62.0	66.1	70.2	74.3	78.4	82.5	86.6	90.7	94.8	99.0	103.1	107.2	111.3	115.4	119.5	123.6	127.7
22	51.7	56.0	60.2	64.4	68.8	72.8	77.1	81.3	85.5	89.7	94.0	98.2	102.4	106.6	110.8	115.1	119.3	123.5	127.7	132.0
23	53.8	58.2	62.5	66.8	71.2	75.5	79.8	84.2	88.5	92.8	97.2	101.5	105.8	110.2	114.5	118.8	123.2	127.5	131.8	136.2
24	56.0	60.4	64.8	69.3	73.7	78.2	82.6	87.1	91.5	96.0	100.4	104.8	109.3	113.7	118.2	122.6	127.1	131.5	136.0	140.4

8'-6"

NUMBER OF SQUARE YARDS OF WALLS AND CEILING IN ROOMS WITH 8'-6" CEILING HEIGHT

Room Size	3	4	5	6	7	8	9	10	11	12	13	14	15	16	17	18	19	20	21	22
3	12.3	14.5	16.7	19.0	21.2	23.4	25.6	27.8	30.1	32.2	34.5	36.7	39.0	41.2	43.4	45.6	47.8	50.1	52.3	54.5
4	14.5	16.8	19.2	21.5	23.8	26.2	28.5	30.8	33.2	35.5	37.8	40.2	42.5	44.8	47.2	49.5	51.8	54.2	56.5	58.8
5	16.7	19.2	21.6	24.1	26.5	29.0	31.4	33.8	36.3	38.7	41.2	43.6	46.1	48.5	51.0	53.4	55.8	58.3	60.7	63.2
6	19.0	21.5	24.1	26.6	29.2	31.7	34.3	36.8	39.4	42.0	44.5	47.1	49.6	52.5	54.7	57.3	59.8	62.4	65.0	67.5
7	21.2	23.8	26.5	29.2	31.8	34.5	37.2	39.8	42.5	45.2	47.8	50.5	53.2	55.8	58.5	61.2	63.8	66.5	69.2	71.8
8	23.4	26.2	29.0	31.7	34.5	37.3	40.1	42.8	45.6	48.4	51.2	54.0	56.7	59.5	62.3	65.1	67.8	70.6	73.4	76.2
9	25.6	28.5	31.4	34.3	37.2	40.1	43.0	45.8	48.7	51.6	54.5	57.4	60.3	63.2	66.1	69.0	71.8	74.7	77.6	80.5
10	27.8	30.8	33.8	36.8	39.8	42.8	45.8	48.8	51.8	54.8	57.8	60.8	63.8	66.8	69.8	72.8	75.8	78.8	81.8	84.8
11	30.1	33.2	36.3	39.4	42.5	45.6	48.7	51.8	55.0	58.1	61.2	64.3	67.4	70.5	73.6	76.7	79.8	83.0	86.1	89.2
12	32.2	35.5	38.7	42.0	45.2	48.4	51.6	54.8	58.1	61.3	64.5	67.7	71.0	74.2	77.4	80.6	83.8	87.0	90.3	93.5
13	34.5	37.8	41.2	44.5	47.8	51.2	54.5	57.8	61.2	64.5	67.8	71.2	74.5	77.8	81.2	84.5	87.8	91.2	94.5	97.8
14	36.7	40.2	43.6	47.1	50.5	54.0	57.4	60.8	64.3	67.7	71.2	74.6	78.1	81.5	85.0	88.4	91.8	95.3	98.7	102.2
15	39.0	42.5	46.1	49.6	53.2	56.7	60.3	63.8	67.4	71.0	74.5	78.1	81.6	85.2	88.7	92.3	95.8	99.4	103.0	106.5
16	41.2	44.8	48.5	52.2	55.8	59.5	63.2	66.8	70.5	74.2	77.8	81.5	85.2	88.8	92.5	96.2	99.8	103.5	107.2	110.8
17	43.4	47.2	51.0	54.7	58.5	62.3	66.1	69.8	73.6	77.4	81.2	85.0	88.7	92.5	96.3	100.1	103.8	107.6	111.4	115.2
18	45.6	49.5	53.4	75.3	61.2	65.1	69.0	72.8	76.7	80.6	84.5	88.4	92.3	96.2	100.1	104.0	107.8	111.7	115.6	119.5
19	47.8	51.8	55.8	59.8	63.8	67.8	71.8	75.8	79.8	83.8	87.8	91.8	95.8	99.8	103.8	107.8	111.8	115.8	119.8	123.8
20	50.1	54.2	58.3	62.4	66.5	70.6	74.7	78.8	83.0	87.1	91.2	95.3	99.4	103.5	107.6	111.7	115.8	120.0	124.1	128.2
21	52.3	56.5	60.7	65.0	69.2	73.4	77.6	81.8	86.1	90.3	94.5	98.7	103.0	107.2	111.4	115.6	119.8	124.1	128.3	132.5
22	54.5	58.8	63.2	67.5	71.8	76.2	80.5	84.8	89.2	93.5	97.8	102.2	106.5	110.8	115.2	119.5	123.8	128.2	132.5	136.8
23	56.7	61.2	65.6	70.1	74.5	79.0	83.4	87.8	92.3	96.7	101.2	105.6	110.1	114.5	119.1	123.4	127.8	132.3	136.7	141.2
24	59.0	63.5	68.1	72.6	77.2	81.7	86.3	90.8	95.4	100.0	104.5	109.1	113.6	118.2	122.7	127.3	131.8	136.4	141.0	145.5

NUMBER OF SQUARE YARDS OF WALLS AND CEILING IN ROOMS WITH 9'-0" CEILING HEIGHT

Room Size	3	4	5	6	7	8	9	10	11	12	13	14	15	16	17	18	19	20	21	22
3	13.0	15.3	17.6	20.0	22.3	24.6	27.0	29.3	31.6	34.0	36.3	38.6	41.0	43.3	45.6	48.0	50.3	52.6	55.0	57.3
4	15.3	17.7	20.2	22.6	25.1	27.5	30.0	32.4	34.8	37.3	39.7	42.2	44.6	47.1	49.5	52.0	54.4	56.8	59.3	61.7
5	17.6	20.2	22.7	25.3	27.8	30.4	33.0	35.5	38.1	40.6	43.2	45.7	48.3	50.8	53.4	56.0	58.5	61.1	63.6	66.2
6	20.0	22.6	25.3	28.0	30.6	33.3	36.0	38.6	41.3	44.0	46.6	49.3	52.0	54.6	57.3	60.0	63.5	65.4	68.0	70.6
7	22.3	25.1	27.8	30.6	33.4	36.2	39.0	41.7	44.5	47.3	50.1	52.8	55.6	58.4	61.2	64.0	66.7	69.5	72.3	75.1
8	24.6	27.5	30.4	33.3	36.2	39.1	42.0	44.8	47.7	50.6	53.5	56.4	59.3	62.2	65.1	68.0	70.8	73.7	77.6	79.5
9	27.0	30.0	33.0	36.0	39.0	42.0	45.0	48.0	51.0	54.0	57.0	60.0	63.0	66.0	69.0	72.0	75.0	78.0	81.0	84.0
10	29.3	32.4	35.5	38.6	41.7	44.8	48.0	52.0	54.2	57.3	60.4	63.5	66.6	69.7	72.8	76.0	79.1	82.2	85.3	88.4
11	31.6	34.8	38.1	41.3	44.5	47.7	51.0	54.2	57.4	60.6	63.8	67.1	70.3	73.5	76.7	80.0	83.2	86.4	89.6	92.8
12	34.0	37.3	40.6	44.0	47.3	50.6	54.0	57.3	60.6	64.0	67.3	70.6	74.0	77.3	80.6	84.0	87.3	90.6	94.0	97.3
13	36.3	39.7	43.2	46.6	50.1	53.5	57.0	60.4	63.8	67.3	70.7	74.2	77.6	81.1	84.5	88.0	91.4	94.8	98.3	101.7
14	38.6	42.2	45.7	49.3	52.8	56.4	60.0	63.5	67.1	70.6	74.2	77.7	81.3	84.8	88.4	92.0	95.5	99.1	102.6	106.2
15	41.0	44.6	48.3	52.0	55.6	59.3	63.0	66.6	70.3	74.0	77.6	81.3	85.0	88.6	92.3	96.0	99.6	103.3	107.0	111.6
16	43.3	47.1	50.8	54.6	58.4	62.2	66.0	69.7	73.5	77.3	81.1	84.8	88.6	92.4	96.2	100.0	103.7	107.5	111.3	115.1
17	45.6	49.5	53.4	57.3	61.2	65.1	69.0	72.8	76.7	80.6	84.5	88.4	92.3	96.2	100.4	104.0	107.8	111.7	115.6	119.5
18	48.0	52.0	56.0	60.0	64.0	68.0	72.0	76.0	80.0	84.0	88.0	92.0	96.0	100.0	104.0	108.0	112.0	116.0	120.0	124.0
19	50.3	54.4	58.5	62.6	66.7	70.8	75.0	79.1	83.2	87.3	91.4	95.5	99.6	103.7	107.8	112.0	116.1	120.2	124.3	128.4
20	52.6	58.6	61.1	65.3	69.5	73.7	78.0	82.2	86.4	90.6	94.8	99.1	103.3	107.5	111.7	116.0	120.2	124.4	128.6	132.8
21	55.0	59.3	63.6	68.0	72.3	76.6	81.0	85.3	89.6	94.0	98.3	102.6	107.0	111.3	115.6	120.0	124.3	128.6	134.3	137.3
22	57.3	61.7	66.2	70.6	75.1	79.5	84.0	88.4	92.8	97.3	101.7	106.2	111.6	115.1	121.5	124.0	128.4	132.8	137.3	141.7
23	59.6	64.2	68.7	73.3	77.8	82.4	87.0	91.5	95.1	100.6	105.2	109.7	114.3	118.8	123.4	128.0	132.5	137.0	141.6	146.2
24	62.0	66.6	71.3	76.0	80.6	85.3	90.0	94.6	99.3	104.0	108.6	113.3	118.0	122.6	127.3	132.0	136.6	141.3	146.0	150.6

9'-6"

NUMBER OF SQUARE YARDS OF WALLS AND CEILING IN ROOMS WITH 9'-6" CEILING HEIGHT

Room Size	3	4	5	6	7	8	9	10	11	12	13	14	15	16	17	18	19	20	21	22
3	13.6	16.1	18.5	21.0	23.4	25.8	28.3	30.7	33.2	35.6	38.1	40.5	43.0	45.4	47.8	50.3	52.7	55.2	57.6	60.1
4	16.1	18.6	21.2	23.7	26.3	28.8	31.4	34.0	36.5	39.1	41.6	44.2	46.7	49.3	51.8	54.4	57.0	59.5	62.1	64.6
5	18.5	21.2	23.8	26.5	29.2	31.8	34.5	37.2	39.8	42.5	45.2	47.8	50.5	53.2	55.8	58.5	61.2	63.8	66.5	69.2
6	21.0	23.7	26.5	29.3	32.1	34.8	37.6	40.4	43.2	46.0	48.7	51.5	54.3	57.1	59.8	62.6	65.4	68.2	71.0	73.7
7	23.4	26.3	29.2	32.1	35.0	37.8	40.7	43.6	46.5	49.4	52.3	55.2	58.1	61.0	63.8	66.7	69.6	72.5	75.4	78.3
8	25.8	28.8	31.8	34.8	37.9	40.8	43.8	46.8	49.8	52.8	55.8	58.8	61.8	64.8	67.8	70.8	73.8	76.8	79.8	82.8
9	28.3	31.4	34.5	37.6	40.7	43.8	47.0	50.1	53.2	56.3	59.4	62.5	65.6	68.7	71.8	75.0	78.1	81.2	84.3	87.4
10	30.7	34.0	37.2	40.4	43.6	46.8	50.1	53.3	56.5	59.7	63.0	66.2	69.4	72.6	75.8	79.1	82.3	85.5	88.7	92.0
11	33.2	36.5	39.8	43.2	46.5	49.8	53.2	56.5	59.8	63.2	66.5	69.8	73.2	76.5	79.8	83.2	86.5	89.8	93.2	96.5
12	35.6	39.1	42.5	46.0	49.4	52.8	56.3	59.7	63.2	66.5	70.1	73.5	77.0	80.4	83.8	87.3	90.7	94.2	97.6	101.1
13	38.1	41.6	45.2	48.7	52.3	55.8	59.4	63.0	66.5	70.1	73.6	77.2	80.7	84.3	87.8	91.4	95.0	98.5	102.1	105.6
14	40.5	44.2	47.8	51.5	55.2	58.8	62.5	66.2	69.8	73.5	77.2	80.8	84.5	88.2	91.8	95.5	99.2	102.8	106.5	110.2
15	43.0	46.7	50.5	54.3	58.1	61.8	61.8	65.6	69.4	73.2	80.7	84.5	88.3	92.1	95.8	99.6	103.4	107.2	111.0	114.7
16	45.4	49.3	53.2	57.1	61.0	64.8	68.7	72.6	76.5	80.4	84.3	88.2	92.1	96.0	99.8	103.7	107.6	111.5	115.4	119.3
17	47.8	51.8	55.8	59.8	63.8	67.8	71.8	75.8	79.8	83.8	87.8	91.8	95.8	99.8	103.8	107.8	111.8	115.8	119.8	123.8
18	50.3	54.4	68.5	62.6	66.7	70.8	75.0	79.1	83.2	87.3	91.4	95.5	99.6	103.7	107.8	112.0	116.1	120.2	124.3	128.4
19	52.7	57.0	61.2	65.4	69.6	73.8	78.1	82.3	86.5	90.7	95.0	99.0	103.4	107.6	111.8	116.1	120.3	124.5	128.7	133.0
20	55.2	59.5	63.8	68.2	72.5	76.8	81.2	85.5	89.8	94.2	98.5	102.8	107.2	111.5	115.8	120.2	124.5	128.8	133.2	137.5
21	57.6	62.1	66.5	71.0	75.4	79.8	84.3	88.7	93.2	97.6	102.1	106.5	111.0	115.4	119.8	124.3	128.7	133.2	137.6	142.1
22	60.1	64.6	69.2	73.7	78.3	82.8	87.4	92.0	96.5	101.1	105.6	110.2	114.7	119.3	123.8	128.4	133.0	137.5	142.1	146.6
23	62.5	67.2	71.8	76.5	81.2	85.8	90.5	95.2	99.8	104.5	109.2	113.8	118.5	123.2	127.8	132.5	137.2	141.8	146.5	151.2
24	65.0	69.7	74.5	79.3	84.1	88.8	93.6	98.4	103.2	108.0	112.7	117.5	122.3	137.1	131.8	136.6	141.4	146.2	151.0	155.7

10'-0"

NUMBER OF SQUARE YARDS OF WALLS AND CEILING IN ROOMS WITH 10'-0" CEILING HEIGHT

Room Size	3	4	5	6	7	8	9	10	11	12	13	14	15	16	17	18	19	20	21	22
3	14.3	16.8	19.4	22.0	24.5	27.1	29.6	32.2	34.7	37.3	39.8	42.4	45.0	47.5	50.1	52.6	55.2	57.7	60.3	62.8
4	16.8	19.5	22.2	24.8	27.5	30.2	32.8	35.5	38.2	40.8	43.5	46.2	48.8	51.5	54.2	56.8	59.5	62.2	64.8	67.5
5	19.4	22.2	25.0	27.7	30.5	33.3	36.1	38.8	41.6	44.4	47.2	50.0	52.7	55.5	58.3	61.1	63.8	66.6	69.4	72.2
6	22.0	24.8	27.7	30.6	33.5	36.4	39.3	42.2	45.1	48.0	50.8	53.7	56.6	59.5	62.4	65.3	68.2	71.1	74.0	76.8
7	24.5	27.5	30.5	33.5	36.5	39.5	42.5	45.5	48.5	51.5	54.5	57.5	60.5	63.5	66.5	69.5	72.5	75.5	78.5	81.5
8	27.1	30.2	33.3	36.4	39.5	42.6	45.7	48.8	52.0	55.1	58.2	61.3	64.4	67.5	70.6	73.7	76.8	80.0	83.1	86.2
9	29.6	32.8	36.1	39.3	42.5	45.7	49.0	52.2	55.4	58.6	61.8	65.1	68.3	71.5	74.7	78.0	81.2	84.4	87.6	90.8
10	32.3	35.5	38.8	42.2	45.5	48.8	52.2	55.5	58.8	62.2	65.5	68.8	72.2	75.5	78.8	82.2	85.5	88.8	92.2	95.5
11	34.7	38.2	41.6	45.1	45.1	48.5	52.0	55.4	58.8	62.2	69.2	72.6	76.1	79.5	83.0	86.4	89.8	93.3	96.7	100.2
12	37.3	40.8	44.4	48.0	51.5	55.1	58.6	62.2	65.7	69.3	72.8	76.4	80.0	83.5	87.1	90.6	94.2	97.7	101.3	104.8
13	39.8	43.5	47.2	50.8	54.5	58.2	61.8	65.5	69.2	72.8	76.5	80.2	83.8	87.5	91.2	94.8	98.5	102.2	105.8	109.5
14	42.4	46.2	50.0	53.7	57.5	61.3	65.1	68.8	72.6	76.4	80.2	84.0	87.7	91.5	95.3	99.1	102.8	106.6	110.4	114.2
15	45.0	48.8	52.7	56.6	60.5	64.4	68.3	72.2	76.1	80.0	83.8	87.7	91.6	95.5	99.4	103.3	107.2	111.1	115.0	118.8
16	47.5	51.5	55.5	59.5	63.5	67.5	71.5	75.5	79.5	83.5	87.5	91.5	95.5	99.5	103.5	107.5	111.5	115.5	119.5	123.5
17	50.1	54.2	58.3	62.4	66.5	70.6	74.7	78.8	83.0	87.1	91.2	95.3	99.4	103.5	107.6	111.7	115.8	120.0	124.1	128.2
18	52.6	56.8	61.1	65.3	69.5	73.7	78.0	82.2	86.4	90.6	94.8	99.1	103.3	107.5	111.7	116.0	120.2	124.4	128.6	132.8
19	55.2	59.5	63.8	68.2	72.5	76.8	81.2	85.5	89.8	94.2	98.5	102.8	107.2	111.5	115.8	120.2	124.5	128.8	133.2	137.5
20	57.7	62.2	66.6	71.1	75.5	80.0	84.4	88.8	93.3	97.7	102.2	106.6	111.1	115.5	120.0	124.4	128.8	133.3	137.7	142.2
21	60.3	64.8	69.4	74.0	78.5	83.1	87.6	92.2	96.7	101.3	105.8	110.4	115.0	119.5	124.1	128.6	133.2	137.7	142.3	146.8
22	62.8	67.5	72.2	76.8	81.5	86.2	90.8	95.5	100.2	104.8	109.5	114.2	118.8	123.5	128.2	132.8	137.5	142.2	146.8	151.5
23	65.4	70.2	75.0	79.7	84.5	89.3	94.1	98.8	103.6	108.4	113.2	118.0	122.7	127.5	132.3	137.1	141.8	146.6	151.4	156.2
24	68.0	72.8	77.7	82.6	87.5	92.4	97.3	102.2	107.1	112.1	114.8	121.8	126.7	131.5	136.4	141.3	146.2	151.0	156.0	160.8

NUMBER OF SQUARE YARDS OF WALLS AND CEILING IN ROOMS WITH 10'-6" CEILING HEIGHT

Room Size	3	4	5	6	7	8	9	10	11	12	13	14	15	16	17	18	19	20	21	22
3	15.0	17.6	20.3	23.0	25.6	28.3	31.0	33.6	36.3	39.0	41.6	44.3	47.0	49.6	52.3	55.0	57.6	60.3	63.0	65.5
4	17.6	20.4	23.2	26.0	28.7	31.5	34.3	37.1	39.8	42.6	45.4	48.2	51.0	53.7	56.5	59.3	62.1	64.8	67.6	70.4
5	20.3	23.2	26.1	29.0	31.8	34.7	37.6	40.5	43.4	46.3	49.2	52.1	55.0	57.8	60.7	63.6	66.5	69.4	72.3	75.2
6	23.0	26.0	29.0	32.0	35.0	38.0	41.0	44.0	47.0	50.0	53.0	56.0	59.0	62.0	65.0	68.0	71.0	74.0	77.0	80.0
7	25.6	28.7	31.8	35.0	38.1	41.2	44.3	47.4	50.5	53.6	56.7	59.8	63.0	66.1	69.2	72.3	75.4	78.5	81.6	84.7
8	28.3	31.5	34.7	38.0	41.2	44.4	47.6	50.8	54.1	57.3	60.6	63.7	67.0	70.2	73.4	76.6	79.8	83.1	86.3	89.5
9	31.0	34.3	37.6	41.0	44.3	47.6	51.0	54.3	57.6	61.0	64.3	67.6	71.0	74.3	77.6	81.0	84.3	87.6	91.0	94.3
10	33.6	37.1	40.5	44.0	47.4	50.8	54.3	57.7	61.2	64.6	68.1	71.5	75.0	78.4	81.8	85.3	88.7	92.2	95.6	99.1
11	36.3	39.8	43.4	47.0	50.5	54.1	57.6	61.2	64.7	68.3	71.8	75.4	79.0	82.5	86.1	89.6	93.2	96.7	100.3	103.8
12	39.0	42.6	46.3	50.0	53.6	57.3	61.0	64.6	68.3	72.0	75.6	79.3	83.0	86.6	90.3	94.0	97.6	101.3	105.0	108.6
13	41.6	45.4	49.2	53.0	56.7	60.5	64.3	68.1	71.8	75.6	79.4	83.2	87.0	90.7	94.5	98.3	102.1	105.8	109.6	113.4
14	44.3	48.2	52.1	56.0	59.8	63.7	67.6	71.5	75.4	79.3	83.2	87.1	91.0	94.8	98.7	102.6	106.5	110.4	114.3	118.2
15	47.0	51.0	55.0	59.0	63.0	67.0	71.0	75.0	79.0	83.0	87.0	91.0	95.0	99.0	103.0	107.0	111.0	115.0	119.0	123.0
16	49.6	53.7	57.8	62.0	66.1	70.2	74.3	78.4	82.5	86.6	90.7	94.8	99.0	103.1	107.2	111.3	115.4	119.5	123.6	127.7
17	52.3	56.5	60.7	65.0	69.2	73.4	77.6	81.8	86.1	90.3	94.5	98.7	103.0	107.2	111.4	115.6	119.8	124.1	128.3	132.5
18	55.0	59.3	63.6	68.0	72.3	76.6	81.0	85.3	89.6	94.0	98.3	102.6	107.0	111.3	115.6	120.0	124.3	128.6	133.0	137.3
19	57.6	62.1	66.5	71.0	75.4	79.8	84.3	88.7	93.2	97.6	102.1	106.5	111.0	115.4	119.8	124.3	128.7	133.2	137.6	142.1
20	60.3	64.8	69.4	74.0	78.5	83.1	87.6	92.2	96.7	101.3	105.8	110.4	115.0	119.5	124.1	128.6	133.2	137.7	142.3	146.8
21	63.0	67.6	72.3	77.0	81.6	86.3	91.0	95.6	100.3	105.0	109.6	114.3	119.0	123.6	128.3	133.0	137.6	142.3	147.0	151.6
22	55.6	70.4	75.2	80.0	84.7	89.5	94.3	99.1	103.8	108.6	113.4	118.2	123.0	127.7	132.5	137.0	142.1	146.8	151.6	156.4
23	68.3	73.2	78.1	83.0	87.8	92.7	97.6	102.5	107.4	112.3	117.2	122.1	127.0	131.8	136.7	141.6	146.5	151.4	156.3	161.2
24	71.0	76.0	81.0	86.0	91.0	96.0	101.0	106.0	111.0	116.0	121.0	126.0	131.0	136.0	141.0	146.0	151.0	156.0	161.0	166.0

11'-0"

NUMBER OF SQUARE YARDS OF WALLS AND CEILING IN ROOMS WITH 11'-0" CEILING HEIGHT

Room Size	3	4	5	6	7	8	9	10	11	12	13	14	15	16	17	18	19	20	21	22
3	15.6	18.4	21.2	24.0	26.7	29.5	32.3	35.1	37.8	40.6	43.4	46.2	49.0	51.7	54.5	57.3	60.1	62.8	65.6	68.4
4	18.4	21.3	24.2	27.1	30.0	32.8	35.7	38.6	41.5	44.4	47.3	50.2	53.1	56.0	58.8	61.7	64.6	67.5	70.4	73.6
5	21.2	24.2	27.2	30.2	33.2	36.2	39.2	42.2	45.2	48.2	51.2	55.2	57.2	60.2	63.2	66.2	69.2	72.2	75.2	78.2
6	24.0	27.1	30.2	33.3	36.4	39.5	42.6	45.7	48.8	52.0	55.1	58.2	61.3	64.4	67.5	70.6	73.7	76.8	80.0	83.1
7	26.7	30.0	33.2	36.4	39.6	42.8	46.1	49.3	52.5	55.7	59.0	62.2	65.4	68.6	71.8	75.1	78.3	81.5	84.7	88.0
8	29.5	32.8	36.2	39.5	42.8	46.2	49.5	52.8	56.2	59.5	62.8	66.2	69.5	72.8	76.2	79.5	82.8	86.2	89.5	92.8
9	32.3	35.7	39.2	42.6	46.1	49.5	53.0	56.4	59.8	63.3	66.7	70.2	73.6	77.1	80.5	84.0	87.4	90.8	94.3	96.7
10	35.1	38.6	42.2	45.7	49.3	52.8	56.4	60.0	63.5	67.1	70.6	74.2	77.7	81.3	84.8	88.4	92.0	95.5	99.1	102.6
11	37.8	41.5	45.2	48.8	52.5	56.2	59.8	63.5	67.2	70.8	74.5	78.2	81.8	85.5	89.2	92.8	96.5	100.2	103.8	107.5
12	40.6	44.4	48.2	52.0	55.7	59.5	63.3	67.1	70.8	74.6	78.4	82.2	86.0	89.7	93.5	97.3	101.1	104.8	108.6	112.4
13	43.4	47.3	51.2	55.1	59.0	62.8	66.7	70.6	74.5	78.4	82.3	86.2	90.1	94.0	97.8	101.7	105.6	109.5	113.4	117.3
14	46.2	50.2	54.2	58.2	62.2	66.2	70.2	74.2	78.2	82.2	86.2	90.2	94.2	98.2	102.2	106.2	110.2	114.2	118.2	122.2
15	49.0	53.1	57.2	61.3	65.4	69.5	73.6	77.7	81.8	86.0	90.1	94.2	98.3	102.4	106.5	110.6	114.7	118.8	123.0	127.1
16	51.7	56.0	60.2	64.4	68.6	72.8	77.1	81.3	85.5	89.7	94.0	98.2	102.4	106.6	110.8	115.1	119.3	123.5	127.7	132.0
17	54.5	58.8	63.2	67.5	71.8	76.2	80.5	84.8	89.2	93.5	97.8	102.2	106.5	110.8	115.2	119.5	123.8	128.2	132.5	136.8
18	57.3	61.7	66.2	70.5	75.1	79.5	84.0	88.4	92.8	97.3	101.7	106.2	110.6	115.1	119.5	124.0	128.4	132.8	137.3	141.7
19	60.1	64.6	69.2	73.7	78.3	82.8	87.4	92.0	96.5	101.1	105.6	110.2	114.7	119.3	123.8	128.4	133.0	137.5	142.1	146.6
20	62.8	67.5	72.2	76.8	81.5	86.2	90.8	95.5	100.2	104.8	109.5	114.2	118.8	123.5	128.2	132.8	137.5	142.2	146.8	151.5
21	65.6	70.4	75.2	80.0	84.7	89.5	94.3	99.1	103.8	108.6	113.4	118.2	123.0	127.7	132.5	137.3	142.1	146.8	151.6	156.4
22	68.4	73.3	78.2	83.1	88.0	92.8	97.7	102.6	107.5	112.4	117.3	122.2	127.1	132.0	136.8	141.7	146.6	151.5	156.4	161.3

12'-0"

NUMBER OF SQUARE YARDS OF WALL AND CEILING IN ROOMS WITH 12'-0" CEILING HEIGHT

Room Size	3	4	5	6	7	8	9	10	11	12	13	14	15	16	17	18	19	20	21	22
3	17.0	20.0	23.0	26.0	29.0	32.0	35.0	38.0	41.0	44.0	47.0	50.0	53.0	56.0	59.0	62.0	65.0	68.0	71.0	74.0
4	20.0	23.1	26.2	29.3	32.4	35.5	38.6	41.7	44.8	48.0	51.1	54.2	57.3	60.4	63.5	66.6	69.7	72.8	76.0	79.1
5	23.0	26.2	29.4	32.6	35.8	39.1	42.3	45.5	48.7	52.0	55.2	58.4	61.6	64.8	68.1	71.3	74.5	77.7	81.0	84.2
6	26.0	29.3	32.6	36.0	39.3	42.6	46.0	49.3	52.6	56.0	59.3	62.6	66.0	69.3	72.6	76.0	79.3	82.6	86.0	89.3
7	29.0	32.4	35.8	39.3	42.7	46.2	49.6	53.1	56.5	60.0	63.4	66.8	70.3	73.7	77.2	80.6	84.4	87.5	91.0	94.4
8	32.0	35.5	39.1	42.6	46.2	49.7	53.3	56.8	60.4	64.0	67.5	71.1	74.6	78.2	81.7	85.3	88.8	92.4	96.0	99.5
9	35.0	38.6	42.3	46.0	49.0	53.3	57.0	60.6	64.3	68.0	71.6	75.3	79.0	82.6	86.3	90.0	93.6	97.3	101.0	104.6
10	38.0	41.7	45.5	49.3	53.0	56.8	60.6	64.4	68.2	72.0	75.7	79.5	83.3	87.1	90.8	94.6	98.4	102.2	106.0	109.7
11	41.0	44.8	48.7	52.6	56.5	60.4	64.3	68.2	72.1	76.0	79.8	83.7	87.6	91.5	95.4	99.3	103.2	107.1	111.0	114.8
12	44.0	48.0	52.0	56.0	60.0	64.0	68.0	72.0	75.0	80.0	84.0	88.0	92.0	96.0	100.0	104.0	108.0	112.0	116.0	120.0
13	47.0	51.1	55.2	59.3	63.4	67.5	71.6	75.7	79.8	84.0	88.1	92.2	96.3	100.4	104.5	108.6	112.7	116.8	121.0	125.0
14	50.0	54.2	58.4	62.6	66.8	71.1	75.3	79.5	83.7	88.0	92.2	96.4	100.6	104.8	109.1	113.3	117.5	121.7	126.0	130.2
15	53.0	57.3	61.6	66.0	70.3	74.6	79.0	83.3	87.6	92.0	96.3	100.6	105.0	109.3	113.6	118.0	122.3	126.6	131.0	135.3
16	56.0	60.4	64.8	69.3	73.7	78.2	82.6	87.1	91.5	96.0	100.4	104.8	109.3	113.7	118.2	122.6	127.1	131.5	136.0	140.4
17	59.0	63.5	68.1	72.6	77.2	81.7	86.3	90.8	95.4	100.0	104.5	109.1	113.6	118.2	122.7	127.3	131.8	136.4	141.0	145.5
18	62.0	66.6	71.3	76.0	80.6	85.3	90.0	94.6	99.3	104.0	108.6	113.3	118.0	122.6	127.3	132.0	136.6	141.3	146.0	150.6
19	65.0	69.7	74.5	79.3	84.1	88.8	93.6	98.4	103.2	108.0	112.7	117.5	122.3	127.1	131.8	136.6	141.4	146.2	151.0	155.7
20	68.0	72.8	77.7	82.6	87.5	92.4	97.3	102.2	107.1	112.0	116.8	121.7	126.6	131.5	136.4	141.3	146.2	151.1	156.0	160.8
21	71.0	76.0	81.0	86.0	91.0	96.0	101.0	106.0	111.0	116.0	121.0	126.0	131.0	136.0	141.0	146.0	151.0	156.0	161.0	166.0
22	74.0	79.1	84.2	89.3	94.4	99.5	104.6	109.7	114.8	120.0	125.1	130.2	135.3	140.4	145.5	150.6	155.7	160.8	166.0	171.1

FUNCTIONS OF NUMBERS, 1 TO 500

No.	Square	Cube
1	1	1
2	4	8
3	9	27
4	16	64
5	25	125
6	36	216
7	49	343
8	64	512
9	81	729
10	100	1000
11	121	1331
12	144	1728
13	169	2197
14	196	2744
15	225	3375
16	256	4096
17	289	4913
18	324	5832
19	361	6859
20	400	8000
21	441	9261
22	484	10648
23	529	12167
24	576	13824
25	625	15625
26	676	17576
27	729	19683
28	784	21952
29	841	24389
30	900	27000
31	961	29791
32	1024	32768
33	1089	35937
34	1156	39304
35	1225	42875
36	1296	46656
37	1369	50653
38	1444	54872
39	1521	59319
40	1600	64000
41	1681	68921
42	1764	74088
43	1849	79507
44	1936	85184
45	2025	91125
46	2116	97336
47	2209	103823
48	2304	110592
49	2401	117649
50	2500	125000
51	2601	132651
52	2704	140608
53	2809	148877
54	2916	157464
55	3025	166375
56	3136	175616
57	3249	185193
58	3364	195112
59	3481	205379
60	3600	216000
61	3721	226981
62	3844	238328
63	3969	250047

FUNCTIONS OF NUMBERS, 1 TO 500

No.	Square	Cube
64	4096	262144
65	4225	274625
66	4356	287496
67	4489	300763
68	4624	314432
69	4761	328509
70	4900	343000
71	5041	357911
72	5184	373248
73	5329	389017
74	5476	405224
75	5625	421875
76	5776	438976
77	5929	456533
78	6084	474552
79	6241	493039
80	6400	512000
81	6561	531441
82	6724	551368
83	6889	571787
84	7056	592704
85	7225	614125
86	7396	636056
87	7569	658503
88	7744	681472
89	7921	704969
90	8100	729000
91	8281	753571
92	8464	778688
93	8649	804357
94	8836	830584
95	9025	857375
96	9216	884736
97	9409	912673
98	9604	941192
99	9801	970299
100	10000	1000000
101	10201	1030301
102	10404	1061208
103	10609	1092727
104	10816	1124864
105	11025	1157625
106	11236	1191016
107	11449	1225043
108	11664	1259712
109	11881	1295029
110	12100	1331000
111	12321	1367631
112	12544	1404928
113	12769	1442697
114	12996	1481544
115	13225	1520875
116	13456	1560896
117	13689	1601613
118	13924	1643032
119	14161	1685159
120	14400	1728000
121	14641	1771561
122	14884	1815848
123	15129	1860867
124	15376	1906624
125	15625	1953125
126	15876	2000376

FUNCTIONS OF NUMBERS, 1 TO 500

No.	Square	Cube
127	16129	2048383
128	16384	2097152
129	16641	2146689
130	16900	2197000
131	17161	2248091
132	17242	2299969
133	17689	2352637
134	17956	2406104
135	18225	2460375
136	18496	2515456
137	18769	2571353
138	19044	2628072
139	19321	2685619
140	19600	2744000
141	19881	2803221
142	20164	2863288
143	20449	2924207
144	20736	2985984
145	21025	3048625
146	21316	3112136
147	21609	3176523
148	21904	3241792
149	22201	3307949
150	22500	3375000
151	22801	3442951
152	23104	3511808
153	23409	3581577
154	23716	3652264
155	24025	3723875
156	24336	3766416
157	24649	3869893
158	24964	3944312
159	25281	4019679
160	25600	4096000
161	25921	4173281
162	26244	4251528
163	26569	4330747
164	26896	4410944
165	27225	4492125
166	27556	4574296
167	27889	4657463
168	28224	4741632
169	28561	4826809
170	28900	4913000
171	29241	5000211
172	29584	5088448
173	29929	5177717
174	30276	5268024
175	30625	5359375
176	30976	5451776
177	31329	5545233
178	31684	5639762
179	32041	5735339
180	32400	5832000
181	32761	5929741
182	33124	6028568
183	33489	6128487
184	33856	6229504
185	34225	6331625
186	34596	6434856
187	34969	6539203
188	35344	6644672
189	35721	6751269

FUNCTIONS OF NUMBERS, 1 TO 500

No.	Square	Cube
190	36100	6859000
191	36481	6967871
192	36864	7077888
193	37249	7189057
194	37636	7301384
195	38025	7414875
196	38416	7529536
197	38809	7645373
198	39204	7762392
199	39601	7880599
200	40000	8000000
201	40401	8120601
202	40804	8242408
203	41209	8365427
204	41616	8489664
205	42025	8615125
206	42436	8741816
207	42849	8869743
208	43264	8998912
209	43681	9129329
210	44100	9261000
211	44521	9393931
212	44944	9528128
213	45369	9663597
214	45796	9800344
215	46225	9938375
216	46656	10077696
217	47089	10218313
218	47824	10360232
219	47961	10503459
220	48400	10648000
221	48841	10793861
222	49284	10941048
223	49729	11089567
224	50176	11239424
225	50625	11390625
226	51076	11543176
227	51529	11697083
228	51984	11852352
229	52441	12008989
230	52900	12167000
231	53361	12326391
232	53824	12487168
233	54289	12649337
234	54756	12812904
235	55225	12977875
236	55696	13144256
237	56169	13312053
238	56644	13481272
239	57121	13651919
240	57600	13824000
241	58081	13997521
242	58564	14172488
243	59049	14348907
244	59536	14526784
245	60025	14706125
246	60516	14886936
247	61009	15069223
248	61504	15252992
249	62001	15438249
250	62500	15625000
251	63001	15813251
252	63504	16003008

FUNCTIONS OF NUMBERS, 1 TO 500

No.	Square	Cube
253	64009	16194277
254	64516	16387064
255	65025	16581375
256	65536	16777216
257	66049	16974593
258	66564	17173512
259	67081	17373979
260	67600	17576000
261	68121	17779581
262	68644	17984728
263	69169	18191447
264	69696	18399744
265	70225	18609625
266	70756	18821096
267	71289	19034163
268	71824	19248832
269	72361	19465109
270	72900	19683000
271	73441	19902511
272	73984	20123648
273	74529	20346417
274	75076	20570824
275	75625	20796875
276	76176	21024576
277	76729	21253933
278	77284	21484952
279	77841	21717639
280	78400	21952000
281	78961	22188041
282	79524	22425768
283	80089	22665187
284	80656	22906304
285	81225	23149125
286	81796	23393656
287	82369	23639903
288	82944	23887872
289	83521	24137569
290	84100	24389000
291	84681	24642171
292	85264	24897088
293	85849	25153757
294	86436	25412184
295	87025	25672375
296	87616	25934336
297	88209	26198073
298	88804	26463592
299	89401	26730899
300	90000	27000000
301	90601	27270901
302	91204	27543608
303	91809	27818127
304	92416	28094464
305	93025	28372625
306	93636	28652616
307	94249	28934443
308	94864	29218112
309	95481	29503629
310	96100	29791000
311	96721	30080231
312	97344	30371328
313	97969	30664297
314	98596	30959144
315	99225	31255875

FUNCTIONS OF NUMBERS, 1 TO 500

No.	Square	Cube
316	99856	31554496
317	100489	31855013
318	101124	32157432
319	101761	32461759
320	102400	32768000
321	103041	33076161
322	103684	33386248
323	104329	33698267
324	104976	34012224
325	105625	34328125
326	106276	34645976
327	106929	34965783
328	107584	35287552
329	108241	35611289
330	108900	35937000
331	109561	36264691
332	110224	36594368
333	110889	36926037
334	111556	37259704
335	112225	37595375
336	112896	37933056
337	113569	38272753
338	114244	38614472
339	114921	38958219
340	115600	39304000
341	116281	39651821
342	116964	40001688
343	117649	40353607
344	118336	40707584
345	119025	41063625
346	119716	41421736
347	120409	41781923
348	121104	42144192
349	121801	42508549
350	122500	42875000
351	123201	43243551
352	123904	43614208
353	124609	43986977
354	125316	44361864
355	126025	44738875
356	126736	45118016
357	127449	45499293
358	128164	45882712
359	128881	46268279
360	129600	46656000
361	130321	47045881
362	131044	47437928
363	131769	47832147
364	132496	48228544
365	133225	48627125
366	133956	49027896
367	134689	49430863
368	135424	49836032
369	136161	50243409
370	136900	50653000
371	137641	51064811
372	138384	51478848
373	139129	51895117
374	139876	52313624
375	140625	52734375
376	141376	53157376
377	142129	53582633
378	142884	54010152

FUNCTIONS OF NUMBERS, 1 TO 500

No.	Square	Cube
379	143641	54439939
380	144400	54872000
381	145161	55306341
382	145924	55742968
383	146689	56181887
384	147456	56623104
385	148225	57066625
386	148996	57512456
387	149769	57960603
388	150544	58411072
389	151321	58863869
390	152100	59319000
391	152881	59776471
392	153664	60236288
393	154449	60698457
394	155236	61162984
395	156025	61629875
396	156816	62099136
397	157609	62570773
398	158404	63044792
399	159201	63521199
400	160000	64000000
401	160801	64481201
402	161604	64964808
403	162409	65450827
404	163216	65939264
405	164025	66430125
406	164836	66923416
407	165649	67419143
408	166464	67917312
409	167281	68417929
410	168100	68921000
411	168921	69426531
412	169744	69934528
413	170569	70444997
414	171396	70957944
415	172225	71473375
416	173056	71991296
417	173889	72511713
418	174724	73034632
419	175561	73560059
420	176400	74088000
421	177241	74618461
422	178084	71551448
423	178929	75686967
424	179776	76225024
425	180625	76765625
426	181476	77308776
427	182329	77854483
428	183184	78402752
429	184041	78953589
430	184900	79507000
431	185761	80062991
432	186624	80621568
433	187489	81182737
434	188356	81746504
435	189225	82312875
436	190096	82881856
437	190969	83453453
438	191844	84027672
439	192721	84604519

FUNCTIONS OF NUMBERS, 1 TO 500

No.	Square	Cube
440	193600	85184000
441	194481	85766121
442	195364	86350888
443	196249	86938307
444	197136	87528384
445	198025	88121125
446	198916	88716536
447	199809	89314623
448	200704	89915392
449	201601	90518849
450	202500	91125000
451	203401	91733851
452	204304	92345408
453	205209	92959677
454	206116	93576664
455	207025	94196375
456	207936	94818816
457	208849	95443993
458	209764	96071912
459	210681	96702579
460	211600	97336000
461	212521	97972181
462	213444	98611128
463	214369	99252847
464	215296	99897344
465	216225	100544625
466	217156	101194696
467	218089	101847563
468	219024	102503232
469	219961	103161709
470	220900	103823000
471	221841	104487111
472	222784	105154048
473	223729	105823817
474	224676	106496424
475	225625	107171875
476	226576	107850176
477	227529	108531333
478	228484	109215352
479	229441	109902239
480	230400	110592000
481	231361	111284641
482	232324	111980168
483	233289	112678587
484	234256	113379904
485	235225	114084125
486	236196	114791256
487	237169	115501303
488	238144	116214272
489	239121	116930169
490	240100	117649000
491	241081	118370771
492	242064	119095488
493	243049	119823157
494	244036	120553784
495	245025	121287375
496	246016	122023936
497	247009	122763473
498	248004	123505992
499	249001	124251499
500	250000	125000000

AREAS OF PLANE FIGURES

SQUARE: Square the length of one side.

$$A = L^2$$

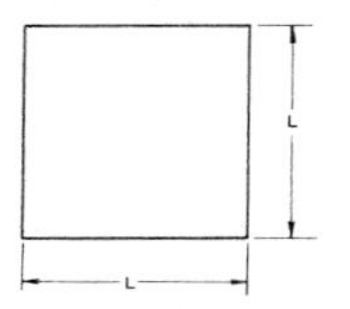

RECTANGLE: Multiply length by width.

$$A = LW$$

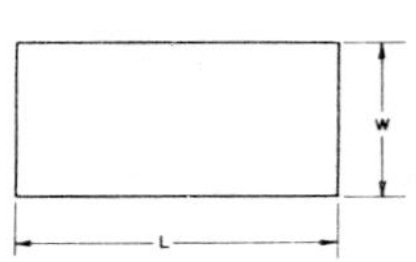

TRIANGLE: Multiply half the altitude by the length of the base.

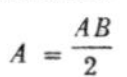

$$A = \frac{AB}{2}$$

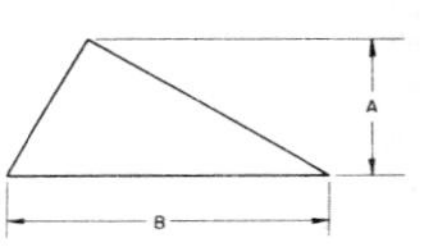

CIRCLE: Square the diameter, and multiply by .7854

$$A = .7854D^2$$

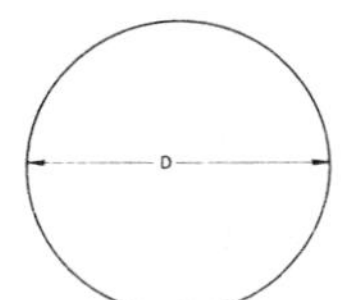

ELLIPSE: Multiply the minor axis by the major axis by .7854.

$$A = .7854Dd$$

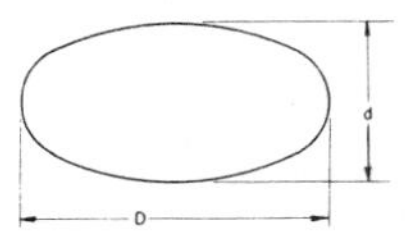

HEXAGON: Square the short diameter and multiply by .866, or square the long diameter and multiply by .6495.

$$A = .866d^2$$

or

$$A = .649D^2$$

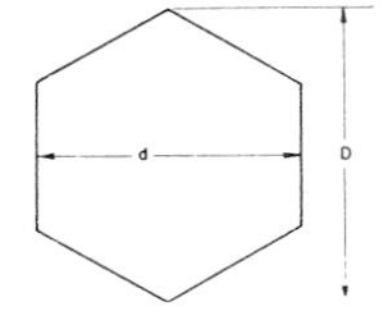

OCTAGON: Square the short diameter and multiply by .828, or square the long diameter and multiply by .707.

$$A = .828d^2$$
$$A = .707D^2$$

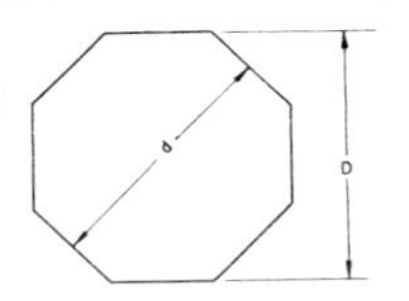

PARALLELOGRAM: Multiply length by perpendicular height.

$$A = LH$$

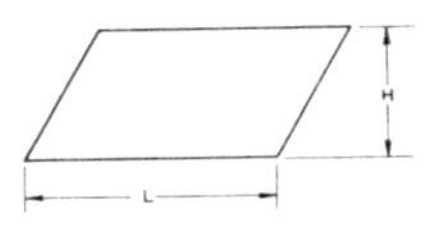

TRAPEZOID: Multiply height by half the sum of the top and bottom bases.

$$A = \frac{H(b-B)}{2}$$

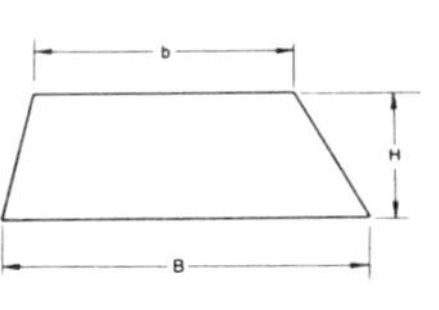

VOLUMES OF TYPICAL SOLIDS

Any prism or cylinder, right or oblique, rectangular or not:

Volume = area of base × altitude
Altitude = distance between parallel bases, measured perpendicular to the bases

When bases are not parallel
Altitude = perpendicular distance from one base to the center of the other

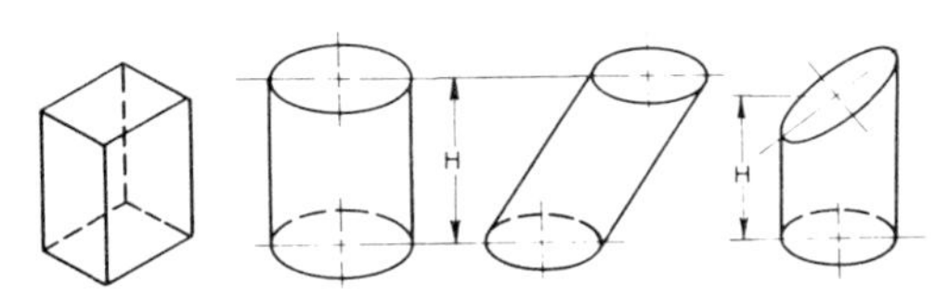

TYPICAL SOLIDS

Any pyramid or cylinder or cone, right or oblique, regular or not:

Volume = area of base × ⅓ altitude
Altitude = distance from base to apex, measured perpendicular to base

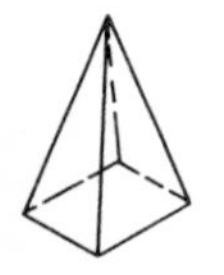

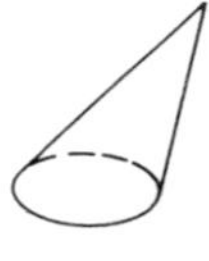
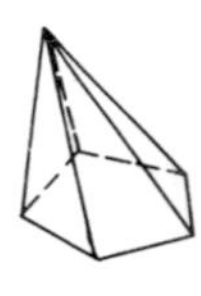

TYPICAL SOLIDS

BETHLEHEM AND CARNEGIE STRUCTURAL SECTIONS AND AMERICAN STANDARD CHANNEL AND BEAM SECTIONS

The following tables give the standard sizes of Bethlehem and Carnegie structural sections, and American standard channel and beam sections. The nomenclature of these tables is as follows:

D = nominal depth, In.
Wt. = weight, lb. per ft.
S = section modulus
d = actual depth, in.
b = flange width, in.
WF = Bethlehem and Carnegie section
B = Bethlehem section
C = Carnegie section
I = American standard section
⊐ = American standard channel section

BETHLEHEM AND CARNEGIE STRUCTURAL SECTIONS

D	Wt.	S	d	b	$\frac{b}{2} + 2$	$\frac{b}{2} + 4\frac{1}{2}$
36 WF	300	1,105.1	36¾	16⅝	10⅜	12⅞
	280	1,031.2	36½	16⅝		
	260	951.1	36¼	16½	10¼	12¾
	250	911.7	36⅛	16½		
	240	873.6	36	16½		
	230	835.5	35⅞	12½		
	194	663.6	36½	12⅛	8⅛	10⅝
	182	621.2	36⅜	12⅛		
	170	579.1	36⅛	12	8	10½
	160	541.0	36	12		
	150	502.9	35⅞	12		
33 WF	240	811.1	33½	15⅞	10	12½
	220	740.6	33¼	15¾	9⅞	12⅜
	210	704.4	33⅛	15¾		
	200	669.6	33	15¾		
	152	486.4	33½	11⅝	7⅞	10⅜
	141	446.8	33¼	11½	7¾	10¼
	132	413.7	33⅛	11½		
	125	385.1	33	11½		
30 WF	210	649.9	30⅜	15⅛	9⅝	12⅛
	200	617.6	30¼	15⅛		
	190	586.1	30⅛	15	9½	12
	180	555.2	30	15		
	172	528.2	29⅞	15		
	132	579.7	30¼	10½	7¼	9¾
	124	354.6	30⅛	10½		
	116	327.9	30	10½		
	108	299.2	29⅞	10½		
27 WF	177	492.8	27¼	14⅛	9⅛	11⅝
	163	452.9	27⅛	14	9	11½
	154	427.8	27	14		
	145	402.9	26⅞	14		
	114	299.2	27¼	10⅛	7⅛	9⅝
	106	277.2	27⅛	10	7	9½
	98	255.3	27	10		
	91	233.2	26⅞	10		
24 WF	160	413.5	24¾	14⅛	9⅛	11⅝
	150	385.5	24½	14⅛		
	140	358.6	24⅜	14	9	11½
	130	330.7	24¼	14		
	120	299.1	24¼	12⅛	8⅛	10⅝
	110	274.4	24¼	12	8	10½
	100	248.9	24	12		
	94	220.9	24¼	9	6½	9
	87	204.3	24⅛	9		
	80	185.8	24	9		
	74	170.4	23⅞	9		

BETHLEHEM AND CARNEGIE STRUCTURAL SECTIO

(Continued)

D	Wt.	S	d	b	$\frac{b}{2} + 2$
21 WF	142	317.2	21½	13⅜	8⅝
	132	294.8	21¼	13⅜	
	122	272.5	21⅛	13	8½
	112	249.6	21	13	
	103	213.1	21¼	9⅛	6⅝
	96	197.6	21⅛	9	6½
	89	182.8	21	9	
	82	168.0	20⅞	9	
	73	150.7	21¼	8¼	6⅛
	68	139.9	21⅛	8¼	
	63	128.0	21	8¼	
	59	119.3	20⅞	8¼	
18 WF	124	239.0	18⅝	11⅞	8
	114	220.1	18½	11⅞	
	105	202.2	18⅜	11¾	7⅞
	96	184.4	18⅛	11¾	
	85	155.1	18⅜	8⅞	6½
	77	141.7	18⅛	8¾	6⅜
	70	128.2	18	8¾	
	64	117.0	17⅞	8¾	
	55	98.2	18⅛	7½	5¾
	50	89.0	18	7½	
	47	82.3	17⅞	7½	
16 WF	114	197.4	16⅝	11⅝	7⅞
	105	181.7	16½	11⅝	
	98	166.1	16⅜	11½	7¾
	88	151.3	16⅛	11½	
	78	127.8	16⅜	8⅝	6⅜
	71	115.9	16⅛	8½	6¼
	64	104.2	16	8½	
	58	94.1	15⅞	8½	
	50	80.7	16¼	7⅛	5⅝
	45	72.4	16⅛	7	5½
	40	64.4	16	7	
	36	56.3	15⅞	7	
14 WF	426	707.4	18¾	16¾	10⅜
	412	682.1	18½	16⅝	
	398	656.9	18¼	16⅝	
	384	632.2	18⅛	16½	10¼
	370	608.1	18	16½	
	356	383.6	17¾	16⅜	
	342	559.4	17½	16⅜	
	328	535.8	17⅜	16¼	10⅛
	314	511.9	17¼	16¼	
	300	488.2	17	16⅛	
	287	465.5	16¾	16⅛	
	273	442.0	16⅝	16⅛	
	264	427.4	16½	16	10
	255	412.0	16⅜	16	

BETHLEHEM AND CARNEGIE STRUCTURAL SECTIONS
(*Continued*)

D	Wt.	S	d	b	$\frac{b}{2}+2$	$\frac{b}{2}+4\frac{1}{2}$
14 WF	246	397.4	16¼	16	10	12½
	237	382.2	16⅛	15⅞		
	228	367.8	16	15⅞		
	219	352.6	15⅞	15⅞		
	211	339.2	15¾	15¾	9⅞	12⅜
	202	324.9	15⅝	15¾		
	193	310.0	15½	15¾		
	184	295.8	15⅜	15⅝		
14 WF	176	281.9	15¼	15⅝	9⅞	12⅜
	167	267.3	15⅛	15⅝		
	158	253.4	15	15½	9¾	12¼
	150	240.2	14⅞	15½		
	142	226.7	14¾	15½		
	320	492.8	16¾	16¾	10⅜	12⅞
	136	216.0	14¾	14¾	9⅜	11⅞
	127	202.0	14⅝	14¾		
	119	189.4	14½	14⅝		
	111	176.3	14⅜	14⅝		
	103	163.6	14¼	14⅝		
	95	150.6	14⅛	14½	9¼	11¾
	87	138.1	14	14½		
14 WF	84	130.9	14⅛	12	8	10½
	78	121.1	14	12		
	74	112.3	14¼	10⅛	7⅛	9⅝
	68	103.0	14	10	7	9½
	61	92.0	13⅞	10		
	58	85.0	14	8⅛	6⅛	8⅝
	53	77.8	14	8	6	8½
	48	70.2	13¾	8		
	43	62.7	13⅝	8		
	42	60.7	14¼	6¾	5⅜	7⅞
	38	54.6	14⅛	6¾		
	34	48.5	14	6¾		
	30	41.8	13⅞	6¾		
12 WF	190	263.2	14⅜	12⅝	8⅜	10⅞
	176	242.6	14⅛	12⅝		
	161	222.2	13⅞	12½	8¼	10¾
	147	201.8	13⅝	12½		
	133	182.5	13⅜	12⅜		
	120	163.4	13⅛	12⅜		
	106	144.5	12⅞	12¼	8⅛	10⅝
	99	134.7	12¾	12¼		
	92	125.0	12⅝	12⅛		
	85	115.7	12½	12⅛		
	79	107.1	12⅜	12⅛		
	72	97.5	12¼	12	8	10½
	65	88.0	12⅛	12		

BETHLEHEM AND CARNEGIE STRUCTURAL SECTIONS
(Continued)

D	Wt.	S	d	b	$\frac{b}{2} + 2$	$\frac{b}{2} + 4\frac{1}{2}$
12 WF	64	85.8	12¼	10	7	9½
	58	78.1	12¼	10		
	53	70.7	12	10		
	50	64.7	12¼	8⅛	6⅛	8⅝
	45	58.2	12	8	6	8½
	40	51.9	12	8		
	36	45.9	12¼	6⅝	5⅜	7⅞
	32	40.7	12⅛	6½	5¼	7¾
	28	35.6	12	6½		
	25	30.9	11⅞	6½		
	22	25.3	12¼	4	4	6½
	19	21.4	12⅛	4		
	16½	17.5	12	4		
	14	14.8	11⅞	4		
10 WF	136	154.4	11⅞	10⅝	7⅜	9⅞
	124	139.9	11⅝	10½	7¼	9¾
	112	126.3	11⅜	10⅜		
	100	112.4	11⅛	10⅜		
	89	99.7	10⅞	10¼	7⅛	9⅝
	77	86.1	10⅝	10¼		
	72	80.1	10½	10⅛		
	66	73.7	10⅞	10⅛		
	60	67.1	10¼	10⅛		
	54	60.4	10⅛	10	7	9½
	49	54.6	10	10		
10 WF	45	49.1	10⅛	8	6	8½
	41	44.5	10	8		
	37	39.9	9⅞	8		
	33	35.0	9¾	8		
	29	30.8	10¼	5¾	4⅛	7⅜
	26	27.6	10⅛	5¾		
	23	24.1	10	5¾		
	21	21.5	9⅞	5¾		
	19	18.8	10¼	4	4	6½
	17	16.2	10⅛	4		
	15	13.8	10	4		
	15	10.5	9⅞	4		
8 WF	67	60.4	9	8¼	6⅛	8⅝
	58	52.0	8¾	8¼		
	48	43.2	8½	8⅛		
	40	35.5	8¼	8⅛		
	35	31.1	8⅛	8	6	8½
	33	29.3	8	8		
	31	27.4	8	8		
	27	23.4	8	6½	5¼	7¾
	24	20.8	7⅞	6½		

BETHLEHEM AND CARNEGIE STRUCTURAL SECTIONS
(Continued)

D	Wt.	S	d	b	$\frac{b}{2} + 2$	$\frac{b}{2} + 4\frac{1}{2}$
8 WF	21	18.0	$8\frac{1}{4}$	$5\frac{1}{4}$		
	19	16.0	$8\frac{1}{8}$	$5\frac{1}{4}$	$4\frac{5}{8}$	$7\frac{1}{8}$
	17	14.1	8	$5\frac{1}{4}$		
	15	11.8	$8\frac{1}{8}$	4		
	13	9.9	8	4	4	$6\frac{1}{2}$
	10	7.8	$7\frac{7}{8}$	4		
7 WF	12	8.5	7	$3\frac{1}{2}$	$3\frac{3}{4}$	$6\frac{1}{4}$
6 B	88	59.5	$7\frac{1}{4}$	$10\frac{3}{8}$	$7\frac{1}{4}$	$9\frac{3}{4}$
	80	53.8	7	$10\frac{3}{8}$		
	70	46.9	$6\frac{7}{8}$	$10\frac{1}{4}$	$7\frac{1}{8}$	$9\frac{5}{8}$
	60	40.0	$6\frac{5}{8}$	$10\frac{1}{8}$		
	50	33.2	$6\frac{3}{8}$	10	7	$9\frac{1}{2}$
	40	26.3	$6\frac{1}{8}$	$9\frac{7}{8}$		
6 C	88	54.7	$6\frac{7}{8}$	10		
	80	49.5	$6\frac{5}{8}$	10	7	$9\frac{1}{2}$
	70	43.0	$6\frac{1}{2}$	$9\frac{7}{8}$		
	60	36.7	$6\frac{1}{4}$	$9\frac{3}{4}$	$6\frac{7}{8}$	$9\frac{3}{8}$
	50	30.4	6	$9\frac{5}{8}$		
	40	24.2	$5\frac{3}{4}$	$9\frac{1}{2}$	$6\frac{3}{4}$	$9\frac{1}{4}$
6 WF	27.5	18.0	$6\frac{1}{4}$	$6\frac{1}{8}$	$5\frac{1}{8}$	$7\frac{5}{8}$
	25	16.4	$6\frac{1}{4}$	6		
	22.5	14.8	$6\frac{1}{8}$	6		
	20	13.1	6	6	5	$7\frac{1}{2}$
	18	11.5	$5\frac{7}{8}$	6		
	16	10.1	$6\frac{1}{4}$	4	4	$6\frac{1}{2}$
	15.5	9.7	$5\frac{3}{4}$	6	5	$7\frac{1}{2}$
6 WF	12	7.24	6	4	4	$6\frac{1}{2}$
	10	5.9	6	3	$3\frac{1}{2}$	6
	8.5	5.07	$5\frac{7}{8}$	4	4	$6\frac{1}{2}$
5 WF	18.9	9.5	5	5		
	18.5	9.94	$5\frac{1}{8}$	5	$4\frac{1}{2}$	7
	16	8.53	5	5		
	13.5	7.02	5	5		
4 C	13.8	5.3	4	4	4	$6\frac{1}{2}$
4 B	13.8	5.45	$4\frac{1}{8}$	4		
	10	4.16	4	4	4	$6\frac{1}{2}$
	7.5	3.13	$3\frac{7}{8}$	4		

AMERICAN STANDARD CHANNEL SECTIONS*

D	Wt.	S	d	b
18 ⊏	58	74.5	18	4 1/4
	51.9	69.1	18	4 1/8
	45.8	63.7	18	4
	42.7	61.0	18	4
15 ⊏	55	57.2	15	3 7/8
	50	53.6	15	3 3/4
	45	49.8	15	3 5/8
	40	46.2	15	3 1/2
	35	42.5	15	3 3/8
	33.9	41.7	15	3 3/8
12 ⊏	40	32.8	12	3 3/8
	35	29.8	12	3 1/4
	30	26.9	12	3 1/8
	25	23.9	12	3
	20.7	21.4	12	3
10 ⊏	35	23.0	10	3 1/8
	30	20.6	10	3
	25	18.1	10	2 7/8
	20	15.7	10	2 3/4
	15.3	13.4	10	2 5/8
9 ⊏	25	15.7	9	2 3/4
	20	13.5	9	2 5/8
	15	11.3	9	2 1/2
	13.4	10.5	9	2 3/8
8 ⊏	21.25	11.9	8	2 5/8
	18.75	10.9	8	2 1/2
	16.25	9.9	8	2 3/8
	13.75	9.0	8	2 3/4
	11.5	8.1	8	2 1/4
7 ⊏	19.75	9.4	7	2 1/2
	17.25	8.6	7	2 3/8
	14.75	7.7	7	2 1/4
	12.25	6.9	7	2 1/4
	9.8	6.0	7	2 1/8
6 ⊏	15.5	6.5	6	2 1/4
	13.0	5.8	6	2 1/8
	10.5	5.0	6	2
	8.2	4.3	6	1 7/8
5 ⊏	11.5	4.1	5	2
	9.0	3.5	5	1 7/8
	6.7	3.0	5	1 3/4
4 ⊏	7.25	2.3	4	1 3/4
	6.25	2.1	4	1 5/8
	5.4	1.9	4	1 5/8
3 ⊏	6.0	1.4	3	1 5/8
	5.0	1.2	3	1 1/2
	4.1	1.1	3	1 3/8

DECIMAL EQUIVALENTS OF FRACTIONS

Inches	Decimal of an Inch	Inches	Decimal of an Inch
1/64	.01562	11/64	.17187
1/32	.03125	3/16	.1875
3/64	.04687	1/5	.2
1/20	.05	13/64	.20312
1/16	.0625	7/32	.21875
1/13	.0769	15/64	.23437
5/64	.07812	1/4	.25
1/12	.0833	17/64	.26562
1/11	.0909	9/32	.28125
3/32	.09375	19/64	.296875
1/10	.10	5/16	.3125
7/64	.10937	21/64	.32812
1/9	.111	1/3	.333
1/8	.125	11/32	.34375
9/64	.14063	23/64	.35937
1/7	.1429	3/8	.375
5/32	.15625	25/64	.39062
1/6	.1667	13/32	.40625

Continued on page 165

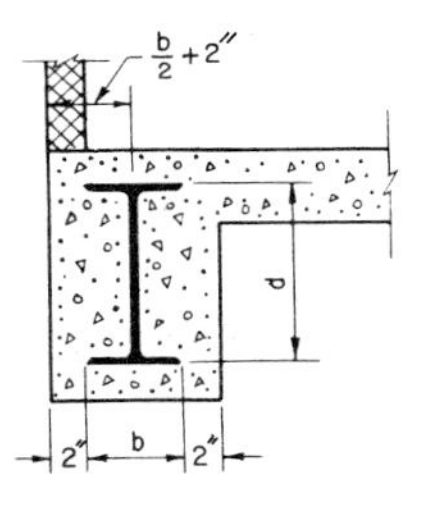

INTERIOR DETAIL

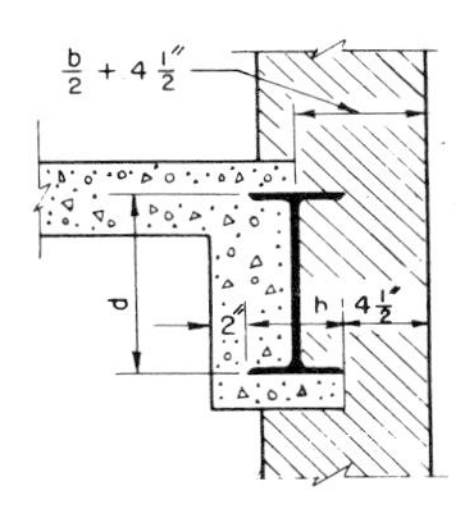

SPANDREL DETAIL

BETHLEHEM AND CARNEGIE STRUCTURAL SECTIONS

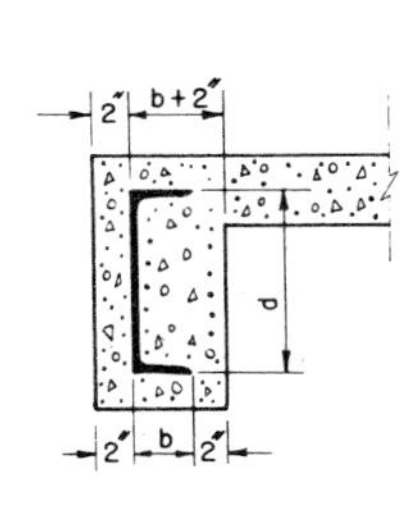

INTERIOR DETAIL

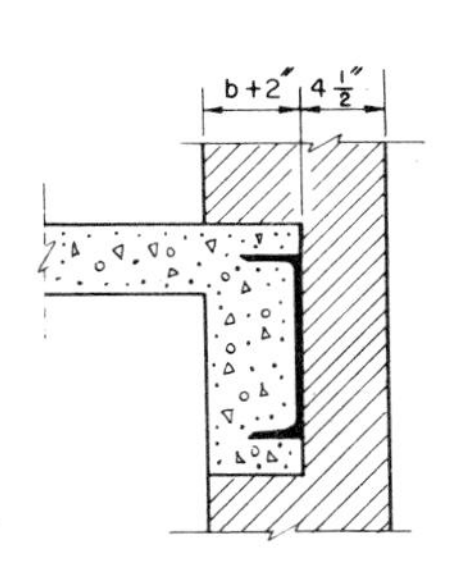

SPANDREL DETAIL

AMERICAN STANDARD CHANNEL SECTIONS

AMERICAN STANDARD BEAM SECTIONS

D	Wt.	S	d	b	$\frac{b}{2}+2$	$\frac{b}{2}+4\frac{1}{2}$
24 I	120	250.9	24	8	6	$8\frac{1}{2}$
	115	245.0	24	8		
	110	239.1	24	$7\frac{7}{8}$		
	105.9	234.3	24	$7\frac{7}{8}$		
	100	197.6	24	$7\frac{1}{4}$	$5\frac{5}{8}$	$8\frac{1}{8}$
	95	191.8	24	$7\frac{1}{8}$		
	90	185.8	24	$7\frac{1}{8}$		
	85	180.0	24	$7\frac{1}{8}$		
	79.9	173.9	24	7	$5\frac{1}{2}$	8
20 I	100	164.8	20	$7\frac{1}{4}$	$5\frac{5}{8}$	$8\frac{1}{8}$
	95	160.0	20	$7\frac{1}{4}$		
	90	155.0	20	$7\frac{1}{8}$		
	85	150.2	20	7	$5\frac{1}{2}$	8
	81.4	146.6	20	7		
	75.0	126.3	20	$6\frac{3}{8}$	$5\frac{1}{4}$	$7\frac{3}{4}$
	70.0	121.4	20	$6\frac{3}{8}$		
	65.4	116.9	20	$6\frac{1}{4}$	$5\frac{1}{8}$	$7\frac{5}{8}$
18 I	70	101.9	18	$6\frac{1}{4}$	$5\frac{1}{8}$	$7\frac{5}{8}$
	65	97.5	18	$6\frac{1}{8}$		
	60	93.1	18	$6\frac{1}{8}$		
	54.7	88.4	18	6	5	$7\frac{1}{2}$
15 I	75	91.6	15	$6\frac{1}{4}$	$5\frac{1}{8}$	$7\frac{5}{8}$
	70	87.9	15	$6\frac{1}{8}$		
	65	84.3	15	$6\frac{1}{8}$		
	60.8	81.2	15	6	5	$7\frac{1}{2}$
	55.0	67.8	15	$5\frac{3}{4}$	$4\frac{7}{8}$	$7\frac{3}{8}$
	50.0	64.2	15	$5\frac{5}{8}$		
	45.0	60.5	15	$5\frac{1}{2}$	$4\frac{3}{4}$	$7\frac{1}{4}$
	42.9	58.9	15	$5\frac{1}{2}$		
12 I	55.0	53.2	12	$5\frac{5}{8}$	$4\frac{7}{8}$	$7\frac{1}{8}$
	50	50.3	12	$5\frac{1}{2}$	$4\frac{3}{4}$	$7\frac{1}{4}$
	45	47.3	12	$5\frac{3}{8}$		
	40.8	44.8	12	$5\frac{1}{4}$	$4\frac{5}{8}$	$7\frac{1}{8}$
	35	37.8	12	$5\frac{1}{8}$		
	31.8	36.0	12	5	$4\frac{1}{2}$	7

AMERICAN STANDARD BEAM SECTIONS

(Continued)

D	Wt.	S	d	b	$\frac{b}{2} + 2$	$\frac{b}{2} + 4\frac{1}{2}$
10 I	40	31.6	10	5 1/8	4 5/8	7 1/8
	35	29.2	10	5	4 1/2	7
	30	26.7	10	4 3/4	4 3/8	6 7/8
	25.4	24.4	10	4 5/8		
8 I	25.8	17.0	8	4 1/4	4 1/8	6 5/8
	23.0	16.0	8	4 1/8		
	20.5	15.1	8	4 1/8		
	18.4	14.2	8	4	4	6 1/2
7 I	20.0	12.0	7	3 7/8	4	6 1/2
	17.5	11.1	7	3 3/4	3 7/8	6 3/8
	15.3	10.4	7	3 5/8		
6 I	12.5	7.3	6	3 3/8	3 3/4	6 1/4
5 I	10.0	4.8	5	3	3 1/2	6
4 I	7.7	3.0	4	2 5/8	3 3/8	5 7/8

DECIMAL EQUIVALENTS OF FRACTIONS

Continued from page 162

Inches	Decimal of an Inch	Inches	Decimal of an Inch
27/64	.42187	45/64	.70312
7/16	.4375	23/32	.71875
29/64	.45312	47/64	.73437
15/32	.46875	3/4	.75
31/64	.48437	49/64	.76562
1/2	.50	25/32	.78125
33/64	.51562	51/64	.79687
17/32	.53125	13/16	.8125
35/64	.54687	53/64	.82812
9/16	.5625	27/32	.84375
17/64	.57812	55/64	.85937
19/32	.59375	7/8	.875
39/64	.60937	57/64	.89062
5/8	.625	29/32	.90625
41/64	.64062	59/64	.92187
21/32	.65625	15/16	.9375
43/64	.67187	61/64	.95312
11/16	.6875	31/32	.96875
		1	1.0

WEIGHTS OF STRUCTURAL STEEL ANGLES

Length of Legs, In.	Thickness, In.	Weight per Ft., Lb.	Length of Legs, In.	Thickness, In.	Weight per Ft., Lb.
8 × 8	1/2	26.4	3 1/2 × 3 1/2	1/4	5.8
	5/8	32.7		5/16	7.2
	3/4	38.9		3/8	8.5
	7/8	45.0		7/16	9.8
	1	51.0		1/2	11.1
	1 1/8			5/8	13.6
6 × 6	3/8	14.9	3 × 3	1/4	4.9
	7/16	17.2		5/16	6.1
	1/2	19.6		3/8	7.2
	9/16	21.9		7/16	8.3
	5/8	24.2		1/2	9.4
	3/4	28.7		5/8	11.5
	7/8	33.1			
	1	37.4	2 1/2 × 2 1/2	1/8	2.08
	1 3/16	39.7		3/16	3.07
				1/4	4.10
5 × 5	3/8	12.3		5/16	5.00
	7/16	14.3		3/8	5.90
	1/2	16.2		1/2	7.70
	5/8	20.0			
	3/4	23.0	2 × 2	1/8	1.65
	7/8	27.2		3/16	2.44
	1	30.6		1/4	3.19
				5/16	3.92
4 × 4	1/4	6.6		3/8	4.70
	5/16	8.2		7/16	5.48
	3/8	9.8		1/2	6.00
	7/16	11.3			
	1/2	12.8	1 3/4 × 1 3/4	1/8	1.44
	5/8	15.7		3/16	2.12
	3/4	18.5		1/4	2.77
	13/16	19.9		5/16	3.39
1 1/2 × 1 1/2	1/8	1.23	3 1/2 × 5	5/16	8.7
	3/16	1.80		3/8	10.4
	1/4	2.34		7/16	12.0
	5/16	2.86		1/2	13.6
	3/8	3.35		5/8	16.8
1 1/4 × 1 1/4	1/8	1.01	3 1/2 × 4	1/4	6.2
	3/16	1.48		5/16	7.7
	1/4	1.92		3/8	9.1
	5/16	2.26		7/16	10.6
				1/2	11.9
1 × 1	1/8	.80			
	3/16	1.16	3 × 4	1/4	5.8
	1/4	1.49		5/16	7.2
				3/8	8.5
6 × 8	1/2	23.0		7/16	9.8
	3/4	33.8		1/2	11.1
				5/8	13.6
3 1/2 × 8	1/2	18.7		3/4	16.0

WEIGHTS OF STRUCTURAL STEEL ANGLES (*Continued*)

Length of Legs, In.	Thickness, In.	Weight per Ft., Lb.	Length of Legs, In.	Thickness, In.	Weight per Ft., Lb.
3½ × 7	3/8	13.0	3 × 3½	1/4	5.4
	7/16	15.0		5/16	6.6
	1/2	17.0		3/8	7.9
	5/8	21.0		1/2	10.2
4 × 6	3/8	12.3	2½ × 3	1/4	4.5
	7/16	14.3		5/16	5.6
	1/2	16.2		3/8	6.6
	9/16	18.1		1/2	8.5
	5/8	20.0	2 × 2½	3/16	2.75
	3/4	23.6	1½ × 2	1/8	1.44
	7/8	27.2		3/16	2.12
				1/4	2.77
				5/16	3.39

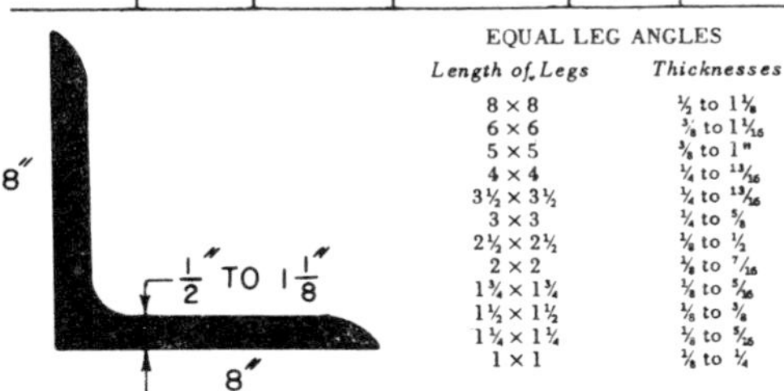

EQUAL LEG ANGLES

Length of Legs	*Thicknesses*
8 × 8	1/2 to 1 1/8
6 × 6	3/8 to 1 1/16
5 × 5	3/8 to 1"
4 × 4	1/4 to 13/16
3½ × 3½	1/4 to 13/16
3 × 3	1/4 to 5/8
2½ × 2½	1/8 to 1/2
2 × 2	1/8 to 7/16
1¾ × 1¾	1/8 to 5/16
1½ × 1½	1/8 to 3/8
1¼ × 1¼	1/8 to 5/16
1 × 1	1/8 to 1/4

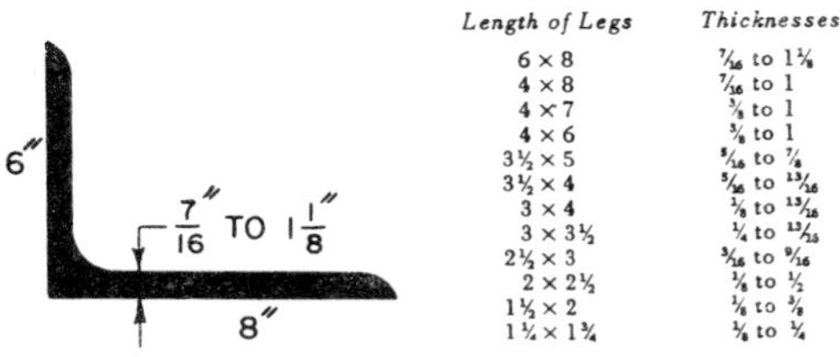

UNEQUAL LEG ANGLES

Length of Legs	*Thicknesses*
6 × 8	7/16 to 1 1/8
4 × 8	7/16 to 1
4 × 7	3/8 to 1
4 × 6	3/8 to 1
3½ × 5	5/16 to 7/8
3½ × 4	5/16 to 13/16
3 × 4	1/4 to 13/16
3 × 3½	1/4 to 13/16
2½ × 3	3/16 to 9/16
2 × 2½	1/8 to 1/2
1½ × 2	1/8 to 3/8
1¼ × 1¾	1/8 to 1/4

STRUCTURAL STEEL

In estimating for structural steel it is best to consider each item separately, as girders, beams, trusses, lintels, columns, and column bases.

GIRDERS: Plate-girder details are often estimated by percentages but usually consist of stiffeners, fillers, and shelf angles. The best method is to figure all members as shown on the plans. An addition of 2½ to 5% is made for rivetheads.

COLUMNS: Columns are estimated by taking the main members as shown on the plans, including bases, caps, splices, and connections. For columns without cover plates, figure 1½ to 3 lb. per lin. ft. for rivetheads and 5 to 8 lb. per lin. ft. where cover plates are used.

For iron column bases figure as shown on the plans, and add 10% for fillets and overrun, except on very large castings when 5 to 7% overrun is sufficient. In roof trusses where a large amount of detail exists figure 15 to 25% as much as the main members.

BEAMS: Beams are figured as shown on the plan, adding all necessary connections obtained from the manufacturer's reference.

STRUCTURAL STEEL

In estimating the probable weight of structural steel for a job, the estimator should determine from the plans the total number of linear feet for each shape by size and weight. However, variations in weight amounting to $2\frac{1}{2}\%$ above or below the nominal weight shown on the plans, are permissible and will occur. The purchaser is charged for the actual weight furnished, provided the weight does not fall outside the permissible variation.

The weight of the steel used for connections should be estimated and priced separately if a detailed estimate is desirable. In estimating the weight of a steel plate having an irregular shape, the weight of the rectangular plate from which the shape is cut should be used. Steel weighs 490 lb/cu. ft. or 0.283 lb/cu. in.

ESTIMATING STEEL LINTELS OVER OPENINGS

Steel lintels over windows and door openings are figured by multiplying the number of lintels over openings by the length of the lintel by the weight of the lintel per linear foot to get the weight in pounds. For example, find the total linear feet of steel angle lintels used over the windows and door openings shown on the following plan. The 3′-4″ windows and the 2′-0″ window require 3 lintels, 4″ x 3″ x 5/16″ over each opening. The two doors require 3 lintels, 6″ x 4″ x 7/16″ over each opening.

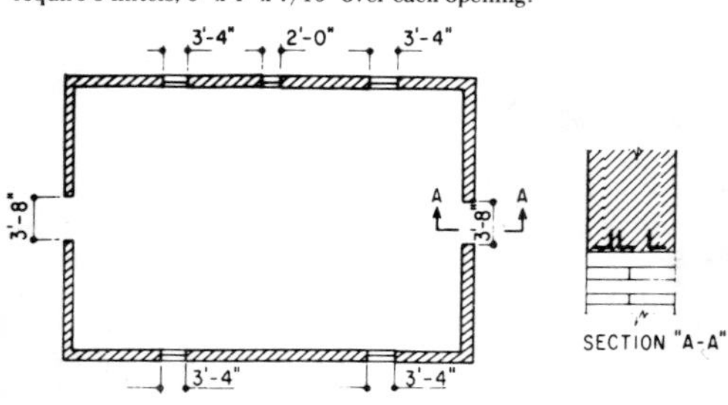

4 windows x 3 lintels x 4′-0″ (lintel length) = 48 ft.
1 window x 3 lintels x 2.67 ft. (lintel length) = 8.01 ft.

total 56.01 ft.

Then,

56.01 ft. x 7.2 (weight of lintel in pounds per foot) = 403.27 lb
2 doors x 3 lintels x 4.33 ft. (length of lintel) = 25.98 ft.

Then,

25.98 x 14.3 (weight of lintel in pounds per foot) = 370.51 lb

Then,

403.27
+ 370.51

773.78 lb 2000 = 0.34 tons

CUBIC INCHES OF VARIOUS BILLET PLATE AND BEARING PLATE SIZES

Billet Plate Size	Cubic Inches	Billet Plate Size	Cubic Inches
14″ x 1¼″ x 1′-2″	245	14″ x 1¼″ x 1′-0″	210
32″ x 3½″ x 1′-0″	1248	33″ x 4″ x 1′-0″	1584
24″ x 2½″ x 1′-0″	720	29″ x 3″ x 2′-5″	2523
14″ x 1″ x 1′-0″	168		
Bearing Plate Size	**Cubic Inches**	**Bearing Plate Size**	**Cubic Inches**
8″ x ¾″ x 0′-8″	48	8″ x 1″ x 1′-0″	96

HOW TO TAKE OFF STEEL BEAMS AND GIRDERS

Find the total length of beams of the same size and multiply the weight per foot of the beam to get the total pounds. Divide the pounds by 2000 to get the total tons.

HOW TO ESTIMATE THE WEIGHT OF BEARING PLATES AND BILLET PLATES

Steel bearing plates are used under the bearing ends of steel beams and girders on foundation walls. Billet plates, anchored on top of reinforced concrete columns and footings, receive the steel columns. The steel columns are usually welded to the billet plates. To estimate the weight of such steel multiply the length of the plate in inches by the width of the plate in inches by the thickness of the plate in inches to get the cubic inches. Multiply the total cubic inches by 0.283 lb. (the weight of the steel per cubic inch). Divide the total weight in pounds by 2000 (1 ton) to find the total tons of steel.

WEIGHTS OF MATERIALS

Exterior Walls and Slabs

Dead Load without Plaster Type of Wall or Slab	Lb. per Sq. Ft.	Lb. per Cu. Ft.
Concrete masonry:		
Cement, stone, sand	....	144
Cement, gravel, sand	....	140
Cement, slag, etc.	....	130
Cement, cinders, etc.	....	108
Hollow gravel concrete block	....	48
Solid cinder concrete block	....	87
Brick masonry:		
Common brick	....	120
Soft brick	....	100
Pressed brick	....	140
4" brick, 6" tile backing	75	
4" brick, 8" tile backing	80	
Hollow tile block	....	62
Face brick, cell type	....	115
4" ordinary brickwork	37	
8" ordinary brickwork	79	
12" ordinary brickwork	115 to 120	
16" ordinary brickwork	155 to 160	

Note: Brick assumed at 4.5 lb. each with ½" mortar joint.

WEIGHTS OF MATERIALS (*Continued*)

Interior Wall and Partitions

Dead Load in Lb. per Sq. Ft. Type of Wall	Lb. per Sq. Ft. Gypsum Plaster Both Sides	Lb. per Sq. Ft. Not Plastered
2" × 4" wood studs and lath	14 to 16	
2" × 3" wood studs and lath	11 to 13	
2" solid plaster	20	
4" solid plaster	32	
4" hollow plaster	22	
2" solid gypsum block	20	9.5
3" solid gypsum block	24	13
3" hollow gypsum block	21	10
4" hollow gypsum block	24	12
5" hollow gypsum block	26	15.5
6" hollow gypsum block	28	17
Split gypsum furring tile, plaster on both sides	12	
2" T.C. hollow block	22 to 25	16
3" T.C. hollow block	27	17
4" T.C. hollow block	28	20 to 22
6" T.C. hollow block	35 to 38	27 to 31
8" T.C. hollow block	42	34 to 36
4" solid brick in P.C. mortar	50	37
8" solid brick in P.C. mortar	90	77
3⅞" glass block		20
3" core-type concrete block		15
4" core-type concrete block	....	20
8" hollow gravel concrete block	....	54
8" hollow cinder concrete block	....	36
8" wall ties	....	33
12" wall ties	....	45

Roofing and Ceiling Finishes

Type of Roof or Ceiling	*Lb. per Sq. Ft.*
Wood shingles	2
Asphalt shingles	3 to 4
Flat clay tile shingles	15
Clay tile roofing (Spanish type)	8.5 to 10
Cement tile	16
Asbestos shingles or siding	2.5 to 3
2" book tile	12
3" book tile	20
Sheet-metal roofing	2
Corrugated iron, No. 20 gage	2
Corrugated asbestos	3 to 4
Wood roofers or sheathing	3
Slag roofing	5
5-ply felt, tar, and gravel roofing	6
4-ply felt, tar, and gravel roofing	5

Roofing and Ceiling Finishes

Type of Roof or Ceiling	*Lb. per Sq. Ft.*
3-ply felt, tar, and gravel roofing	4
3-ply Ready (composition) roofing	1
Slate roofing, $\frac{3}{16}$" and $\frac{1}{4}$" thickness	7 to 9.5
Slate roofing, $\frac{3}{8}$" thickness	14.5
Slate roofing, 3" double-lap $\frac{1}{2}$" thickness	19.5
Skylights—frame and glass	10
Plate glass per inch thick	14
Hung ceilings—M.L. and $\frac{3}{4}$" P.C. plaster	10
Hung ceilings—lime or gypsum plaster	8
Lime or gypsum plaster on walls	5
Lime or gypsum plaster on slabs	6

Flooring and Floor Slabs

Type of Floor	*Lb. per Sq. Ft.*
4" cinder concrete (floor arches and slabs)	36
4" cinder concrete (fill over Fireproof floors)	20
Cinder fill per in. of thickness	5 to 7
Cement finish, 1" thick	12
Terrazzo, $1\frac{1}{2}$" thick	18
Tile and setting bed	15 to 23
Marble and setting bed	25 to 30
Asphalt mastic flooring, $1\frac{1}{2}$" thick	18
$\frac{25}{32}$" or $\frac{7}{8}$" hardwood flooring	4
$\frac{3}{8}$" hardwood flooring	2
$\frac{25}{32}$" or $\frac{7}{8}$" softwood underflooring	3
2" × 4" wood sleepers and fill	10
Oak and longleaf yellow pine	48*
Fir, spruce, hemlock, white pine	30*
3" creosoted wood flooring	15
Linoleum or bonded flooring	4
Patented structural systems:	
Gypsteel plank, 2" thick	12
Sheetrock—Pyrofill, $2\frac{1}{2}$" thick	12
Featherweight nailing concrete, $2\frac{1}{2}$" thick	19
Porete roof slabs, $2\frac{1}{2}$" thick	15
Featherweight channel slabs, $2\frac{3}{4}$" thick	10
Porete channel slabs, $3\frac{1}{2}$" thick	12
Cinder plank, 2" thick	15
Aerocrete lightweight concrete	50 to 80*
Nalecode	75*

*In pounds per cubic foot.

Miscellaneous Weights

Material	*Lb.*
1000 brick, stacked, lb. per cu. ft.	56
Sand or gravel, dry and loose, lb. per cu. ft.	90 to 105
Sand or gravel, packed, lb. per cu. ft.	100 to 120
P.C., sand, and lime mortar, lb. per cu. ft.	103 to 116
4" extra-heavy cast-iron pipe per ft.	13

RESIDENTIAL CONSTRUCTION DATA

Standard dead-load computation for residential floor tier of wood joists, cross bridging, underfloor, hardwood, linoleum, or bonded finish flooring, and also lath and plaster ceiling: 17 lb. per sq. ft. No ceiling, deduct 8 lb. per sq. ft. For a tile floor instead of wood add 25 lb.

INSTANT MULTIPLICATION TABLE*

EXAMPLE: Multiply 13 × 12. Find 13 in first column. Read across to number under 12. Answer is 156.

	1	2	3	4	5	6	7	8	9	10	11	12	13	14
1	1													
2	2	4												
3	3	6	9											
4	4	8	12	16										
5	5	10	15	20	25									
6	6	12	18	24	30	36								
7	7	14	21	28	35	42	49							
8	8	16	24	32	40	48	56	64						
9	9	18	27	36	45	54	63	72	81					
10	10	20	30	40	50	60	70	80	90	100				
11	11	22	33	44	55	66	77	88	99	110	121			
12	12	24	36	48	60	72	84	96	108	120	132	144		
13	13	26	39	52	65	78	91	104	117	130	143	156	169	
14	14	28	42	56	70	84	98	112	126	140	154	168	182	196
15	15	30	45	60	75	90	105	120	135	150	165	180	195	210
16	16	32	48	64	80	96	112	128	144	160	176	192	208	224
17	17	34	51	68	85	102	119	136	153	170	187	204	221	238
18	18	36	54	72	90	108	126	144	162	180	198	216	234	252
19	19	38	57	76	95	114	133	152	171	190	209	228	247	266
20	20	40	60	80	100	120	140	160	180	200	220	240	260	280
21	21	42	63	84	105	126	147	168	189	210	231	252	273	294
22	22	44	66	88	110	132	154	176	198	220	242	264	286	308
23	23	46	69	92	115	138	161	184	207	230	253	276	299	322
24	24	48	72	96	120	144	168	192	216	240	264	288	312	336
25	25	50	75	100	125	150	175	200	225	250	275	300	325	350
26	26	52	78	104	130	156	182	208	234	260	286	312	338	364
27	27	54	81	108	135	162	189	216	243	270	297	324	351	378
28	28	56	84	112	140	168	196	224	252	280	308	336	364	392
29	29	58	87	116	145	174	203	232	261	290	319	348	377	406
30	30	60	90	120	150	180	210	240	270	300	330	360	390	420
31	31	62	93	124	155	186	217	248	279	310	341	372	403	434
32	32	64	96	128	160	192	224	256	288	320	352	384	416	448
33	33	66	99	132	165	198	231	264	297	330	363	396	429	462
34	34	68	102	136	170	204	238	272	306	340	374	408	442	476
35	35	70	105	140	175	210	245	280	315	350	385	420	455	490
36	36	72	108	144	180	216	252	288	324	360	396	432	468	504
37	37	74	111	148	185	222	259	296	333	370	407	444	481	518
38	38	76	114	152	190	228	266	304	342	380	418	456	494	532
39	39	78	117	156	195	234	273	312	351	390	429	468	507	546
40	40	80	120	160	200	240	280	320	360	400	440	480	520	560

*Section of "Instant Calculator." Courtesy of Ira Edward Kornfeld, 2121 Beekman Place, Brooklyn 25, N.Y., publisher/owner of copyright (reprint 25¢, 1954). This includes a complete conversion chart on fractions (½ to 1/99:–98/99 to 3 dec. places)

WEIGHTS OF MATERIALS

Materials	Weight lbs. per cu. ft.
Asphalt-Pavement Composition	100
Bluestone	160
Brick, Dry Pressed	150
Brick, Common and Hard	125
Brickwork in Lime Mortar, average	120
Brickwork in Cement Mortar, average	130
Brickwork, Pressed Brick, thin joints	140
Birch (wood)	48
Cypress (wood)	36
Fir, Douglas	36
Fire Brick	150
Granite	167
Gypsum Block (partition Hollow)	48
Hemlock (wood)	30
Hollow Tile Partition Block	60
Iron, Cast	450
Iron, Wrought	480
Limestone	155–172
Maple (wood)	48
Marble	170–180
Masonry, Squared Granite or Limestone	165
Masonry, Sandstone	150
Mineral Wool	12
Mortar, Hardened	90–100
Oak (wood)	48
Plaster	96
Slate	172–178
Spruce (wood)	30
Steel, structural	489.6
Terra Cotta, solid	120
Terra Cotta, masonry work	80
Tile, solid	120
Yellow Pine (wood)	48

NEW LUMBER STANDARDS

Nominal Size (inches)	Size when Dry or Seasoned, 19% moisture content or under (inches)	Size when Green or unseasoned, over 19% moisture content (inches)
1 x 4	¾ x 3½	25/32 x 3 9/16
1 x 6	¾ x 5½	25/32 x 5 5/8
1 x 8	¾ x 7¼	25/32 x 7½
1 x 10	¾ x 9¼	25/32 x 9½
1 x 12	¾ x 11¼	25/32 x 11½
2 x 4	1½ x 3½	1 9/16 x 3 9/16
2 x 6	1½ x 5½	1 9/16 x 5 5/8
2 x 8	1½ x 7¼	1 9/16 x 7½
2 x 10	1½ x 9¼	1 9/16 x 9½
2 x 12	1½ x 11¼	1 9/16 x 11½
4 x 4	3½ x 3½	3 9/16 x 3 9/16
4 x 6	3½ x 5½	3 9/16 x 5 5/8
4 x 8	3½ x 7¼	3 9/16 x 7½
4 x 10	3½ x 9¼	3 9/16 x 9½
4 x 12	3½ x 11¼	3 9/16 x 11½

DECIMALS OF AN INCH TO MILLIMETERS

Fraction	Decimal Equivalent	Millimeters
9/32	.28125	7.14375
5/16	.3125	7.9375
11/32	.34375	8.73125
⅜	.375	9.525
13/32	.40625	10.31875
7/16	.4375	11.1125
15/32	.46875	11.90625
½	.5	12.7
17/32	.53125	13.49375
9/16	.5625	14.2875
19/32	.59375	15.08125
⅝	.625	15.875
21/32	.65625	16.66875
11/16	.6875	17.4625
23/32	.71875	18.25625
¾	.75	19.05
25/32	.78125	19.84375
13/16	.8125	20.6375
27/32	.84375	21.43125
⅞	.875	22.225
29/32	.90625	23.01875
15/16	.9375	23.8125
31/32	.96875	24.60625
1	1.	25.4

Multiply decimal equivalent by 25.4 to get millimeters.

INCHES TO MILLIMETERS

Inches	Millimeters	Inches	Millimeters
1	25.4	51	1295.4
2	50.8	52	1320.8
3	76.2	53	1346.2
4	101.6	54	1376.6
5	127.0	55	1397.0
6	152.4	56	1422.4
7	177.8	57	1447.8
8	203.2	58	1473.2
9	228.6	59	1498.6
10	254.0	60	1524.0
11	279.4	61	1549.4
12	304.8	62	1574.8
13	330.2	63	1600.2
14	355.6	64	1625.6
15	381.0	65	1651.0
16	406.4	66	1676.4
17	431.8	67	1701.8
18	457.2	68	1727.2
19	482.6	69	1752.6
20	508.0	70	1778.0
21	533.4	71	1803.4
22	558.8	72	1828.8
23	584.2	73	1854.2
24	609.6	74	1879.6
25	635.0	75	1905.0
26	660.4	76	1930.4
27	685.8	77	1955.8
28	711.2	78	1981.2
29	736.8	79	2006.6
30	762.0	80	2032.0
31	787.4	81	2057.4
32	812.8	82	2082.8
33	838.2	83	2108.2
34	863.6	84	2133.6
35	889.0	85	2159.0
36	914.4	86	2184.4
37	939.8	87	2209.8
38	965.2	88	2235.2
39	990.6	89	2260.6
40	1016.0	90	2286.0
41	1041.4	91	2311.4
42	1066.8	92	2336.8
43	1092.2	93	2362.2
44	1117.6	94	2387.6
45	1143.0	95	2413.0
46	1168.4	96	2438.4
47	1193.8	97	2463.8
48	1219.2	98	2489.2
49	1244.6	99	2514.6
50	1270.0	100	2540.0

FIRE RATINGS FOR WALLS AND PARTITIONS

4—HOUR FIRE TEST

8″ Solid Brick
8″ Plain Concrete
8″ Solid Structural Units
8″ Solid Cinder Concrete Blocks
8″ Hollow Clay Tile, 3 cells, plastered both sides
8″ Hollow Concrete Blocks, one-piece cells 1½″ thick, plastered both sides
6″ Solid Reinforced Concrete
6″ Solid Cinder Concrete, plastered both sides
12″ Hollow Concrete Blocks, one-piece webs, 4 cells 1½″ thick, 2 cells in wall thickness

3—HOUR FIRE TEST

8″ Solid Brick
8″ Solid Structural Unit
6″ Plain Concrete
6″ Hollow Concrete Blocks, cells 1½″ thick if unplastered 1¼″ if plastered
6″ Solid Cinder Concrete Blocks
6″ Hollow Clay Tile, 2 cells thick, plastered on room side
5″ Solid Reinforced Concrete
4″ Hollow Gypsum Blocks, plastered on both sides

2—HOUR FIRE TEST

6″ Solid Brick
6″ Solid Structural Units
8″ Hollow Concrete Blocks
6″ Hollow Clay Tile, 2 cells, plastered on one side
4″ Solid, Plain or Reinforced Concrete
4″ Solid Cinder Blocks or 3″ plastered on both sides
4″ Hollow Clay Tile, 2 cells, plastered both sides
3″ Solid Gypsum Blocks, plastered both sides
3″ Solid Gypsum, poured or block

1—HOUR FIRE TEST

4″ Solid or Hollow Brick
4″ Solid Structural Units
3″ Solid Cinder Concrete
3″ Hollow Clay Tile, plastered on both sides
3″ Hollow Gypsum Block
2″ Solid Gypsum Block

LOOK FOR THESE FINE TITLES AT YOUR LOCAL BOOKSTORE OR FILL OUT THIS COUPON

YES. Please send me the following bestselling reference paperbacks:

QTY.	TITLE	AUTHOR	ISBN	PRICE
___	**BUILDER'S VEST POCKET REFERENCE BOOK**	Hornung	0-13-085944-3	$6.95
___	**CONSTRUCTION DRAFTER'S VEST POCKET REFERENCE BOOK**	Hornung	0-13-168815-4	$7.95
___	**ELECTRONICS VEST POCKET REFERENCE BOOK**	Thomas	0-13-252379-5	$6.95
___	**MECHANICS VEST POCKET REFERENCE BOOK**	Wolf & Phelps	0-13-571505-9	$6.95
___	**PIPEFITTER'S AND PLUMBER'S VEST POCKET REFERENCE BOOK**	Bachmann	0-13-676460-6	$6.95
___	**ROOFING, SIDING AND PAINTING CONTRACTOR'S VEST POCKET REFERENCE BOOK**	Hornung	0-13-782475-0	$7.95

PRENTICE HALL PRESS, A Division of Simon & Schuster, Inc.., 200 Old Tappan Road, Old Tappan, NJ 07675

I have enclosed my ☐ check ☐ money order. Or, please charge my ☐ MasterCard ☐ VISA. I've included $1.50 postage and handling for the first book and $1.00 each additional. Please add appropriate sales tax.

______________________ Exp. date __________

Name ______________________________

Address ______________________________

City ______________________________

State ____________________ Zip __________